The Pointless Book

Elias Verdan

Heinrich Wilson Publishing

Contents

INTRODUCTION

You bought a book called The Pointless Book.

Take a moment to consider that. You spent money, or at the very least you spent time, to open a book that tells you in its own title that it has nothing to give you. No solutions. No revelations. No moment of clarity where everything falls into place and you close the cover feeling like you understand something you did not understand before. The title warned you. And you opened it anyway.

That says something about you. Something worth paying attention to. Because the fact that you are here, reading these words, despite the title, despite the warning, means that somewhere in your mind there is a voice that does not believe it. A voice that says the title must be a trick. A provocation. A marketing strategy. Surely no one writes a book that is actually pointless. Surely there is a payoff hidden somewhere. A twist at the end. A lesson wrapped in irony. Something.

I am not going to tell you whether that voice is right or wrong. Not yet. That is for you to decide after the last page, not the first.

What I will tell you is what this book is about and what it is going to do.

This book asks questions. Not obscure questions. Not questions that require a degree in physics or philosophy to understand. They are the kind of questions that almost every human being has asked at some point, usually late at night, usually alone, usually in that strange quiet space where the mind stops being busy and starts being honest.

What was before the universe? What does it expand into? Where did all the matter come from? What is nothing? Why do we need answers to everything? Is there a limit to what we can know? What happens when logic itself breaks down? Why does the question "why" stop working at a certain depth?

You have probably asked some of these yourself. And you have probably noticed that when you did, nobody gave you a satisfying answer. Someone may have given you a scientific explanation. Someone may have given you a religious one. Someone may have said something about quantum mechanics or God or the multiverse or the limits of human understanding. And for a moment, it may have felt like enough. The discomfort faded. The question receded. You moved on.

But the question did not move on. It is still there. Sitting exactly where you left it. Unanswered. Waiting.

This book is going to pick those questions back up. One by one. And for each one, I am going to do something that might seem contradictory given the title. I am going to try to answer them. Genuinely try. Not with handwaving or mysticism or vague gestures toward the unknowable. With science. With quantum theory. With philosophy. With logic. With every tool available to a mind that wants to understand the nature of reality.

And then I am going to show you why it is pointless.

Not because the tools are bad. They are extraordinary. Not because the theories are wrong. Many of them are among the greatest achievements of human thought. But because every tool, every theory, every framework, when pushed far enough, arrives at the same place. A wall. A silence. A point where the explanation ends and the question remains, untouched, exactly as it was before the explanation began.

I am going to show you this wall. From different directions, using different questions, through different disciplines. And each time, the wall will look the same. Not because I am repeating myself. Because the wall is the same. It does not matter which question you ask or which approach you use. The boundary is identical. The silence on the other side is identical. The feeling of reaching it is identical.

That feeling is the point of this book. Not the questions. Not the theories. Not the explanations. The feeling of reaching the edge of what can be known and standing there with nothing to hold onto. Most books try to protect you from that feeling. They give you an answer, any answer, to fill the space. A conclusion. A lesson. A take-away. Something to carry with you so that you do not have to leave empty-handed.

This book will not do that. This book will walk you to the edge and leave you there. And by the end, you will understand something that no answer could have given you. Not a fact. Not a theory. A recognition. A shift in how you see the questions themselves.

I hope you enjoy the journey. I mean that sincerely, even though the destination does not exist. There is something in the act of trying that has value, even when the trying leads nowhere. There is something in standing at the edge that changes how you see everything behind you. And there is something in reading a pointless book that, against all logic, against all expectation, might turn out to be one of the more meaningful things you do this year.

Or it might not. It might be exactly what the title says. Pointless. Chapters of questions that go nowhere, followed by a conclusion that concludes nothing.

I cannot promise you it will be worth it. I can only promise you that the questions are real, the attempts are honest, and the walls are exactly where I say they are.

You are here. The book is open. There are questions ahead that cannot be answered.

Let us try anyway.

1

What Was Before the Universe

What was before the universe?

It is one of the first questions a child asks and one of the last questions a physicist can attempt to address. It feels simple. It feels like it should have an answer. Everything we know has a before. You had a life before today. Today had a yesterday. Yesterday had a last week, a last year, a last century. We move backward through time as naturally as breathing. So it seems perfectly logical to keep going. To rewind past the formation of Earth, past the first stars, past the earliest light, all the way back to the beginning. And then to ask: what was before that?

The problem is not that we lack information. The problem is that the question itself may not work.

Modern cosmology places the origin of the observable universe at approximately 13.8 billion years ago. This is the point described by the

Big Bang model, a moment of extraordinary density and temperature from which space, energy, and matter emerged. It is not an explosion in the traditional sense. There was no empty room in which something detonated. The Big Bang was not an event that happened inside space. It was the event that created space. And with it, according to general relativity, it also created time.

This is where the question breaks.

If time began with the universe, then the word "before" has no reference point. Before is a time-dependent concept. It requires a timeline that already exists. Asking what came before time is like asking what is south of the South Pole. You can construct the sentence. The grammar works. The words are real. But the sentence does not point to anything. It is a dead end dressed up as a direction.

Most people, when they hear this explanation, reject it instinctively. It does not feel right. Our entire experience of reality is built on sequences. On cause and effect. On the assumption that every event was preceded by another event. To be told that there was no "before" feels less like an answer and more like an evasion. It feels like the scientist is dodging the question rather than addressing it.

But the scientist is not dodging. The scientist is pointing out that the question contains a hidden assumption that does not hold. It assumes time is universal and infinite. It assumes time exists independently of the universe. And if that assumption is wrong, then the question was never valid to begin with.

That does not make the question disappear, though. It just makes it harder.

Some physicists have tried to work around this by proposing models in which time, or something like time, does exist before the Big Bang. One of these is the cyclic model, which suggests our universe is one in a potentially infinite series of expansions and contractions. In this framework, the Big Bang was not a beginning. It was a transition. A bounce from a previous contracting phase into a new expanding one. This would mean there was a "before" after all, just not the kind of before we imagine.

But the cyclic model does not solve the deeper problem. It pushes it back. If this universe was preceded by another, what preceded that one? And the one before that? You either end up with an infinite chain of universes, which raises its own unanswerable questions, or you eventually hit a true beginning, which returns you to the same wall. What was before that?

Another attempt comes from quantum cosmology. In certain models, particularly those built on quantum gravity frameworks, the concept of time dissolves entirely at very small scales. Near the Planck epoch, the earliest measurable moment after the Big Bang, the distinction between space and time becomes blurred. In the Hartle-Hawking no-boundary proposal, the universe does not have a boundary in time. Time, as we understand it, gradually becomes something more like a spatial dimension as you approach the origin. There is no edge. There is no starting point. There is no moment zero.

This is a mathematically elegant idea. It removes the need for a "before" by removing the boundary that would require one. But elegance is not the same as truth. The Hartle-Hawking model is a theoretical proposal. It has not been tested. It cannot be tested with any technology currently available or even conceivable in the near future. It is an explanation, not an answer. It tells us what might be the case, not what is the case.

And this is the pattern that repeats itself throughout this book. An impossible question is asked. Brilliant minds offer explanations. Those explanations are sophisticated, well-reasoned, and internally consistent. But none of them are answers. An answer closes a question. An explanation keeps it open while making it feel less uncomfortable.

The discomfort matters. It is not a flaw in human thinking. It is a feature of the question.

The human mind evolved to solve problems. To find food. To avoid predators. To plan for seasons. Every mechanism in our cognition is geared toward finding patterns, causes, and solutions. When we encounter a question that has no answer, the brain does not accept it. It keeps searching. It keeps reaching. It generates discomfort because discomfort is the brain's way of saying this problem is not resolved yet, keep working on it.

But some problems are not unresolved. They are unresolvable. The brain does not have a category for that. There is no mental filing cabinet labeled "questions that cannot be answered." So the mind either keeps grinding, or it reaches for something that feels like an answer even if it is not one.

This is where religion enters the picture.

For thousands of years, the most common response to the question of what came before the universe has been God. A creator. A force that existed outside of time and space and chose to bring everything into being. In Christianity, Islam, Judaism, Hinduism, and nearly every other belief system, some version of this answer exists. God was there before everything. God is eternal. God does not need a beginning because God exists beyond the rules that govern everything else.

This explanation has satisfied billions of people across thousands of years. And on the surface, it appears to work. It fills the gap. It provides a cause for the effect. It stops the infinite regression of "but what came before that?" by placing something at the start that does not require a before.

But it does not actually answer the question. It relocates it.

If God created the universe, what created God? The standard response is that God is eternal, uncreated, and exists outside of time. But this is the same structure as the scientific claim that time began with the universe. It is a boundary statement. It says: past this point, the question does not apply. The only difference is the language used to frame it. Science says time did not exist before the Big Bang. Religion says God exists outside of time. Both are saying the same thing in different words: stop asking, the question does not work here.

And neither is an answer.

If God has always existed, then existence without a beginning is possible. But if existence without a beginning is possible, then why does the universe need a creator at all? If something can simply be, without a cause and without a start, then the universe itself could be that something. The argument for God as the uncaused first cause undermines itself the moment it claims that an uncaused existence is possible. It opens a door it cannot close.

If God was born, then who were his parents? If God created himself, then he existed before he existed, which is a logical impossibility. If God emerged from nothing, then something can come from nothing, and again, the universe does not need a creator. Every path leads back to the same wall. The question does not get answered. It just wears a different costume.

This is not an attack on religion. This is not a statement about whether God exists or does not exist. That is a different question for a different book. This is simply an observation that when it comes to answering the question "what was before the universe," the concept of a creator god does not resolve the problem. It mirrors it. It creates an equally unanswerable question and then forbids you from asking it.

And forbidding a question is not the same as answering it.

So where does that leave us?

It leaves us exactly where we started. The question "what was before the universe?" has no answer. Not because we have not found one yet. Not because science needs more time or religion needs more faith. But because the question may be structurally broken. It uses concepts that do not apply to the thing it is asking about. It assumes time is infinite. It assumes cause and effect operate at every level of reality. It assumes "before" is always a valid direction.

None of those assumptions have been confirmed. All of them might be wrong.

The honest position, the position that very few people are comfortable with, is this: we do not know. And we may never know. Not because we are not smart enough. Not because the answer is hidden somewhere waiting to be discovered. But because the question itself might not have a destination. It might be a road that looks real on the map but does not exist in the landscape.

There is a word for what people feel when they sit with this realization for too long. It is not frustration exactly. It is something closer to vertigo. The mental equivalent of looking over the edge of something with no bottom. The mind searches for ground and finds nothing. Not darkness. Not emptiness. Just the absence of anything to stand on.

Most people, when they reach that point, turn back. They return to an explanation that feels solid enough to stand on, whether it is science or religion or philosophy. They pick one and commit to it. Not because it answers the question, but because the alternative is standing on nothing. And the human mind was not built to stand on nothing.

But the question remains.

Quantum mechanics offers one more attempt worth examining. In quantum field theory, the vacuum, what we call empty space, is not actually empty. It is a seething, fluctuating field of virtual particles that pop in and out of existence. Some physicists have proposed that the universe itself could be a quantum fluctuation. A temporary ripple in a deeper quantum vacuum that inflated and became stable enough to persist.

This is a fascinating idea. It suggests the universe did not need a cause in the classical sense. It emerged as a statistical event in a quantum field. No intention. No design. No before. Just probability expressing itself.

But even this does not escape the trap. If the universe emerged from a quantum vacuum, then the quantum vacuum existed before the universe. And what created the quantum vacuum? What are the rules that govern it? Where did those rules come from? The regression continues. You have not answered the question. You have added another layer to it.

Every explanation, no matter how sophisticated, does the same thing. It pushes the mystery back one step. It creates the illusion of progress without reaching a destination. And at the end of every chain of reasoning, the same silent wall stands waiting. The human mind reaches it, presses against it, and finds nothing on the other side.

Not because nothing is there. But because "there" might not exist.

This is the first lesson of this book. Some questions do not have answers. Not yet. Not ever. The question "what was before the universe?" is not a puzzle waiting to be solved. It is a mirror. It shows us the edges of our own thinking. It shows us where language fails, where logic folds, where the mind runs out of road.

And when you reach that edge, you have two choices. You can turn back and pretend the road continues. Or you can stand there and look at the blank space ahead.

Neither choice gives you an answer.

That is the point.

2

What Does the Universe Expand Into

What does the universe expand into?

It is one of the most frequently asked questions in popular science and one of the most frequently misunderstood. The universe is expanding. This is not speculation. It is one of the most well-supported observations in modern cosmology. Galaxies are moving apart from each other. The space between them is stretching. The rate of this expansion has been measured, refined, and confirmed across decades of observation. It is happening right now, in every direction, at every scale beyond the local gravitational pull of galaxy clusters.

So if the universe is expanding, it must be expanding into something. There must be a space outside the universe that it is growing into. A room. A void. A larger container of some kind.

That is the assumption. And that assumption is wrong.

The expansion of the universe is not like a balloon inflating inside a room. It is not like a gas spreading into a vacuum. Those analogies are used constantly in popular science writing, and they are misleading every single time. They create a mental image that does not reflect what is actually happening. The balloon analogy in particular has probably done more damage to public understanding of cosmology than any other comparison in history.

When cosmologists say the universe is expanding, they do not mean it is getting bigger inside a pre-existing space. They mean space itself is stretching. The fabric of the universe, the dimensions in which everything exists, is increasing in scale. There is no outside. There is no room. There is no container. The universe is not expanding into anything because there is no "anything" outside of it for it to expand into.

This is nearly impossible for the human mind to visualize. And that is not a failure of intelligence. It is a limitation of experience.

Every expansion we have ever witnessed in our lives has happened inside a larger space. A fire spreads across a field. A city grows outward into the countryside. A puddle of water widens on a flat surface. In every case, the expanding thing is smaller than the environment it exists within. Our brains have been trained by a lifetime of these observations to assume that expansion requires a surrounding space. It is one of the deepest intuitions we carry, and it does not apply here.

The universe is not an object inside a space. The universe is space. It includes all dimensions, all distances, all positions. When it expands, it is not moving outward into something. It is creating more of itself. The distances between points within it are increasing. But there is no external point from which you could observe this happening. There is no outside perspective. There is no vantage point beyond the edge.

Mathematically, this is described through general relativity. Einstein's field equations allow for a universe that is finite but unbounded, meaning it has no edge, no wall, no border, yet it is not infinite in the way most people imagine. The most common analogy used to illustrate this is the surface of a sphere. If you walk in any direction on the surface of the Earth, you will never reach an edge. You will never fall off. The surface is finite, it has a measurable area, but it has no boundary. Now remove one dimension from your thinking. The surface of the Earth is a two-dimensional analogy for three-dimensional space. The universe may be curved in a similar way, closed in on itself, with no edge and no outside.

But analogies are not answers. They are tools for making the unimaginable feel slightly less alien. The moment you push on them too hard, they collapse. The surface of the Earth exists within a three-dimensional space. You can fly above it. You can look down at it. So the analogy immediately makes you ask: what is the higher-dimensional space that the universe curves through? And the honest answer is that there may not be one. The curvature described in general relativity does not require an external dimension to curve into. It is intrinsic curvature. The mathematics work without a container.

That answer satisfies the equations. It does not satisfy the brain.

The brain wants a picture. It wants a model. It wants to zoom out far enough to see the whole thing from the outside, the way you can look at a globe on a desk. But you cannot zoom out of the universe. There is no desk. There is no room the desk sits in. The universe is the room and the desk and the air around it and the space that contains all of those things. There is nowhere to stand outside of it.

Some theoretical frameworks do suggest the possibility of something beyond our observable universe. The multiverse hypothesis, which arises in certain interpretations of quantum mechanics and in eternal inflation models, proposes that our universe may be one of many. In the eternal inflation picture, the Big Bang that produced our universe was a local event in a much larger, eternally inflating space. Other regions of this space may have undergone their own Big Bangs, producing other universes with potentially different physical constants and laws.

This is a serious theoretical proposal. It is discussed in peer-reviewed physics. It is not fringe science. But it is also not confirmed science. There is currently no observational evidence for other universes. No signal. No trace. No measurement. The multiverse remains a mathematical possibility, not a demonstrated reality. And even if it is real, it does not answer the original question. It expands it.

If our universe exists within a multiverse, what does the multiverse expand into? What contains the container? The regression is identical to the one we encountered in Chapter 1. Every answer becomes a new question. Every frame we place around the picture reveals that there must be a larger frame beyond it. And at the end of the chain, the same blank wall.

String theory and M-theory, which attempt to unify quantum mechanics with gravity, propose additional spatial dimensions beyond the three we experience. In some of these models, our universe is a three-dimensional "brane," a membrane floating in a higher-dimensional space called the "bulk." Other branes may exist alongside ours, potentially giving rise to other universes.

This is elegant and mathematically rich. It gives the mind something to grab onto. A picture. A structure. A higher-dimensional room in which our universe is just one sheet among many. But once again, it is a theoretical framework, not an observed reality. No experiment has detected extra dimensions. No instrument has measured the bulk. And even within these models, the same question resurfaces. What contains the bulk? What is outside the higher-dimensional space? The frame keeps getting bigger, but it never disappears.

There is a pattern forming here, and it is worth naming directly. The human mind works in containers. Every concept we hold is placed inside a larger concept. A city is inside a country. A country is inside a continent. A continent is on a planet. A planet is in a solar system. A solar system is in a galaxy. A galaxy is in a cluster. A cluster is in the observable universe. And then the mind reaches for the next container and finds nothing.

This is not a failure of science. It is a collision between how reality works and how the mind models it. The mind builds nested boxes. Reality does not appear to be nested. At some point, the boxes end. And the mind has no protocol for that.

Consider a simpler version of the problem. Imagine someone asks you: what is outside this room? You say the hallway. What is outside the hallway? The building. What is outside the building? The street. The city. The country. The atmosphere. The solar system. You keep going. Each answer is valid. Each answer works. And at every stage, the question still makes sense because there is always a larger structure surrounding the smaller one.

Now apply that to the universe. What is outside the universe? The question feels identical in structure to every previous version. But it is not. Every previous version worked because the thing being asked about was a subset of a larger system. The universe is not a subset. It is the set. It contains all subsets. The rules of containment that applied at every previous level do not apply here, because there is no higher level. The question has the same grammar as the ones before it, but it does not have the same meaning.

Language does not warn you when this happens. Language lets you construct the question as if it were perfectly valid. The sentence "what is outside the universe?" follows all the rules of grammar and syntax. It sounds meaningful. It feels meaningful. But it may point to nothing at all.

This is one of the central problems this book keeps returning to. Language is a tool built for a specific environment. It was developed to describe the world as experienced by human beings on the surface of a planet, at a particular scale, within a particular range of speeds and energies. It works brilliantly for that purpose. But when you aim it at questions that operate beyond that environment, it does not break down obviously. It continues to function grammatically while failing to function meaningfully. It keeps producing sentences that look like questions but are not.

Some physicists attempt to sidestep this by redefining terms. They say the universe does not expand "into" anything. Instead, the metric of space increases. The distances between objects grow, not because the objects are moving through space, but because the space between them is scaling up. This is a precise and accurate description. It is also one that almost no human brain can turn into a mental image. It is language describing mathematics, not language describing experience. And for most people, it lands as unsatisfying. It sounds like an evasion. It sounds like the physicist is replacing a real question with a technical restatement that avoids giving a real answer.

And in a sense, that is exactly what is happening. Not dishonestly. Not evasively. But structurally. The physicist cannot give a real answer because a real answer would require pointing to something outside the universe, and there is nothing to point to. The best they can do is redescribe the situation in more precise language and hope that the precision reduces the confusion. Sometimes it does. Most of the time, it does not.

There is also a psychological dimension to this question that is rarely discussed. The idea that the universe has no outside is, for many people, subtly terrifying. Not in a dramatic way. Not in the way a horror film is terrifying. But in a quiet, existential way. We are accustomed to the idea that there is always somewhere else. That if things go wrong here, there is a there to escape to. The idea that there is no "there" beyond the universe removes that psychological safety net. It creates a sense of confinement that is entirely abstract but deeply felt.

This is not a scientific concern. It is a human one. But it influences how people engage with the question. Many people who reject the idea that the universe has no outside are not rejecting the physics. They are rejecting the feeling. The claustrophobia of totality. The idea that everything, including them, is contained within something that has no exterior. It is not a comfortable thought. And the human mind is very skilled at discarding uncomfortable thoughts in favor of more bearable ones.

Religion once again offers an apparent escape. If God exists outside the universe, then there is an outside. There is something beyond. The universe is not the final container. God is. But as we have already seen, this does not resolve the problem. It extends it. What contains God? If God exists outside the universe, where is that? What is the nature of the space God occupies? Is it spatial? Is it temporal? If it has properties, those properties need an explanation. If it has no properties, then in what sense does it exist? The container outside the container still needs a container, and the regression continues with no end.

Simulation theory offers another angle. If the universe is a computed simulation running on some form of technology in a "base reality," then the universe does expand into something. It expands within the memory and processing architecture of whatever system is running it.

Our space is data. Our expansion is the allocation of more data. The container is the machine.

This is an interesting idea, and it has been taken seriously by physicists and philosophers. But it changes the question without answering it. If our universe is a simulation, what is the nature of the base reality? Does that reality have edges? Does it expand? What contains it? You have replaced one mystery with another and gained nothing except a new layer of abstraction.

This is the trap. Every framework that attempts to answer "what does the universe expand into?" either denies the validity of the question or replaces it with an equivalent question wearing different clothes. No framework eliminates it. No theory closes it. It remains open, not because we need more data, but because the structure of the question may not allow for closure.

The universe expands. This is measurable. This is real. But what it expands into, if that question even makes sense, is not something we can observe, test, measure, or possibly even think about coherently. We are inside the system asking about the outside of the system, using tools that were built to operate inside the system. It is like asking a fish to describe the air. The fish has no apparatus for it. Not because the air is too complex, but because nothing in the fish's existence has prepared it for a medium it was never designed to perceive.

We are the fish. The question is the air. We can feel that something is there, at the boundary of our understanding. But we cannot describe it. We cannot reach it. We cannot confirm it exists.

And so the question remains. Unanswered. Possibly unanswerable. Sitting exactly where it was when we started, unmoved by every theory, every framework, every model we have aimed at it.

The universe is expanding.

Into what?

We do not know. We may never know. And the discomfort of that silence is something no explanation has managed to fill.

3

Can Something Come From Nothing

Can something come from nothing?

This is perhaps the oldest question in human thought. It predates science. It predates philosophy as a formal discipline. It sits at the foundation of every creation myth, every cosmological model, every religious tradition, and every scientific theory about the origin of the universe. At some point, there was nothing. And then there was something. How?

Or was there ever nothing at all?

The question sounds straightforward, but it contains at least two problems that most people never examine. The first is the assumption that "nothing" is a coherent concept. The second is the assumption that the transition from nothing to something is an event that requires a mechanism. Both of these assumptions feel obvious. Both of them may be wrong.

Let us start with what science says.

The Big Bang model describes the universe emerging from an extraordinarily hot, dense state approximately 13.8 billion years ago. In the first fractions of a second, the fundamental forces separated, particles formed, and the expansion of space began. Within minutes, the lightest elements, hydrogen and helium, were produced through nucleosynthesis. Over billions of years, gravity pulled this material together into stars, and inside those stars, heavier elements were forged. Carbon. Oxygen. Iron. Silicon. The material that makes up planets, oceans, and human bodies was cooked in the cores of dying stars and scattered across space when those stars exploded.

This accounts for where matter came from after the Big Bang. It does not account for where the Big Bang came from.

The total energy of the universe is a subject of ongoing calculation and debate, but one widely discussed possibility is that it is zero. Or very close to zero. The positive energy of matter and radiation may be exactly balanced by the negative energy of gravitational fields. If this is the case, then the universe did not require a net input of energy to come into existence. It is, in energy terms, nothing rearranged into something. A zero split into a positive and a negative that together still equal zero.

This is not a fringe idea. It has been discussed by physicists including Stephen Hawking, Lawrence Krauss, and Alan Guth. It is mathematically plausible within the framework of general relativity and quantum field theory. And it is profoundly strange.

It suggests that the universe is, in a sense, nothing. That everything we see, every galaxy, every star, every atom, is a rearrangement of zero. Not something that emerged from nothing, but nothing that organized itself into the appearance of something. The books you have read. The air you are breathing. The planet beneath your feet. All of it, in total, may add up to exactly zero.

But this raises an immediate objection that is difficult to dismiss. Even if the total energy of the universe is zero, the universe clearly exists. It has structure. It has matter. It has laws. It has time. A zero that has been split into components that interact with each other in complex ways is not the same as an undisturbed zero. A cheque for a million dollars and a debt of a million dollars add up to zero, but the person holding both is not in the same situation as someone holding neither. The mathematics may balance, but the experience does not.

So the claim that the universe came from nothing because its total energy is zero is a mathematical statement, not an experiential one. It describes the ledger. It does not describe the reality of living inside that ledger. And for most people, the question "can something come from nothing?" is not a question about energy accounting. It is a question about existence. Why is there anything at all, rather than absolute, complete, total absence?

Quantum mechanics offers another angle. In quantum field theory, the vacuum state, what we would colloquially call "empty space," is not empty. It is a field with a minimum energy state, and that field fluctuates. Virtual particles appear and disappear constantly, borrowing energy from the vacuum for brief periods before annihilating. These are not theoretical abstractions. Their effects have been measured. The Casimir effect, the Lamb shift, and other observable phenomena confirm that the quantum vacuum is active, dynamic, and far from nothing.

Some physicists have proposed that the universe itself may have originated as a quantum fluctuation. A random event in a pre-existing quantum vacuum that, instead of disappearing as virtual particles normally do, inflated rapidly and became stable. The universe as an accident. A statistical anomaly that persisted.

This is a real hypothesis. It has mathematical backing. It has been published and debated seriously. But it does not answer the question.

If the universe emerged from a quantum vacuum, then the quantum vacuum is not nothing. It is something. It is a field with properties, with energy, with rules governing its behavior. It has mathematical structure. It obeys laws. A quantum vacuum is not the absence of everything. It is a very specific state of a very specific system. Calling it "nothing" is a redefinition of the word that most people would not accept.

And this is where the argument gets tangled in a way that many popular science books fail to acknowledge. When Lawrence Krauss wrote about the universe coming from nothing, the "nothing" he described was a quantum vacuum. When critics pointed out that a quantum vacuum is not nothing, they were correct. When Krauss responded that this is the closest thing to nothing that physics can define, he was also correct. The disagreement is not about physics. It is about the word "nothing." And the word "nothing" is where the entire question collapses.

What do we actually mean by nothing?

If nothing means the absence of matter, then a quantum vacuum qualifies. If nothing means the absence of energy, then it does not, because the vacuum has a minimum energy state. If nothing means the absence of space, then neither the vacuum nor any physical model qualifies, because every model in physics operates within some kind of spatial or mathematical framework. If nothing means the absolute absence of everything, including laws, fields, dimensions, potential, structure, and properties of any kind, then physics has never described nothing. Physics cannot describe nothing. Physics is the study of something. Asking physics to explain nothing is like asking a hammer to describe the absence of nails.

This is not a limitation of current physics. It may be a permanent boundary. The tools of science operate within frameworks. They require variables, constants, equations, and measurable quantities. True nothing, if such a state is even possible, would have none of those things. It would have no framework. No rules. No potential for anything. And if it had no potential for anything, then nothing could ever emerge from it. Which means either true nothing never existed, or the concept itself is incoherent.

Philosophy has wrestled with this longer than science has. The ancient Greek philosopher Parmenides argued that nothing cannot exist, because to think about nothing is to think about something, and therefore nothing, the moment it is conceived, becomes something. This sounds like wordplay. It is not. It is a genuine logical problem. The concept of absolute nothing may be self-defeating. The moment you define it, you give it properties. The moment you give it properties, it is no longer nothing.

Later philosophers tried different approaches. Leibniz asked the question directly: why is there something rather than nothing? He concluded that there must be a necessary being, something that exists by its own nature and requires no external cause. This, for Leibniz, was God. But as we have already seen, this does not resolve the problem. It places a label on the mystery and calls it solved. If God is the necessary being that requires no cause, then the concept of a causeless existence is possible. And if a causeless existence is possible, the universe itself could be that causeless existence. The argument folds in on itself.

Heidegger called it the fundamental question of metaphysics. He believed this was the deepest question human beings could ask. He also believed that most people spend their entire lives avoiding it.

Not because they cannot understand it, but because sitting with it honestly produces a form of anxiety that has no resolution. He called it Angst. Not fear of something specific. Not worry about a problem that can be solved. But a deep, formless unease that arises when the mind confronts the groundlessness of its own existence.

Heidegger was not a physicist. His language is dense, often frustratingly abstract. But on this point, he identified something that science tends to skip over. This is not just an intellectual puzzle. It is an existential one. It does not only ask about the origin of matter. It asks about the foundation of everything, including the person asking. And when that foundation turns out to be absent, or unknowable, or possibly incoherent, the effect is not purely academic. It is personal.

Most people do not sit with this long enough to feel that effect. They encounter it briefly, feel the first tremor of discomfort, and then reach for an explanation that stops the trembling. Some reach for God. Some reach for quantum mechanics. Some reach for the multiverse. Some reach for simulation theory. Each of these provides a surface to stand on. Each of these creates the feeling of resolution. But none of them are answers. They are explanations. And an explanation holds a question at arm's length. An answer buries it.

Simulation theory proposes that our reality is a computed construct, generated by an advanced civilization or system in a more fundamental layer of existence. In this framework, creation is not a paradox at all. The universe did not emerge from an absence. It was built. It was programmed. It was switched on. The matter, the energy, the laws of physics, all of it was written into the code.

This is psychologically satisfying because it mirrors human experience. We create things constantly. A programmer writes code and a virtual world appears. A novelist writes words and characters exist.

Creation feels familiar. So the idea that the universe was created in a similar way by a more powerful version of us feels intuitive.

But it does not resolve anything. It defers. If our universe was created by a civilization in a base reality, then what created that base reality? Where did their matter come from? Where did their laws of physics originate? The problem has not been solved. It has been exported to another address.

Every attempt follows the same pattern. It either redefines nothing to mean something, or invokes a prior something to explain the current something. No framework, scientific, religious, or philosophical, has ever produced a genuine account of absolute absence giving rise to absolute existence. Not because thinkers have not tried. They have tried for millennia. But because the question may be constructed in a way that no answer can satisfy.

Consider the possibility that the question is not hard. Consider the possibility that it is broken.

It assumes that "nothing" is a possible state. It assumes that there was, or could have been, a condition of absolute absence. But what if there was never nothing? What if something has always existed in some form? What if the concept of a time when nothing existed is as meaningless as the concept of a place north of the North Pole?

If something has always existed, then the question disappears. There was no transition because there was never an absence to transition from. But this creates its own problem. If something has always existed, then existence has no beginning. And if existence has no beginning, then an infinite past stretches behind us. An infinite series has no endpoint, so how did we arrive at this particular moment? An infinite number of events would have to have already occurred for us to be here now, and an infinite number, by definition, is never complete.

This is not a scientific objection. It is a logical one. And it has been debated for centuries without resolution. The idea of an infinite past is no more comfortable or coherent than the idea of existence emerging from absence. Both options resist the mind's attempt to make them make sense.

The honest situation is this. There are only a few possible responses. Either something emerged from absolute absence, which seems impossible. Or something emerged from something, which triggers an infinite regression. Or something has always existed, which contradicts our understanding of beginnings. Or the question is malformed and does not function the way we assume it does.

None of these close the book on the problem. Each one leaves a gap where the mind reaches for solid ground and finds air.

The water in the oceans was forged in stellar explosions billions of years ago. The stone beneath your feet was assembled from elements cooked inside dying suns. The helium in the atmosphere was produced in nuclear reactions that occurred minutes after the Big Bang. Science can trace the history of every atom in your body back to specific physical processes. It can tell you where each element was made and how it got here. The chain of explanations is long, detailed, and supported by extraordinary evidence.

And it ends in silence.

Every theory says maybe. Every framework builds a bridge that stops halfway across the gap. We have better language now than we had a thousand years ago. Better mathematics. Better instruments. But the gap has not closed. It is still there, untouched by everything we have aimed at it.

Pointless, you might say. A question that leads nowhere no matter which direction you approach it from. And maybe that is exactly what it is. But if it is truly pointless, it is strange how much it weighs.

4

Does Everything Have a Beginning and an End

Does everything have a beginning and an end?

It seems self-evident. You were born. You will die. The candle is lit. The candle burns out. The day starts. The day ends. Every experience you have ever had reinforces the idea that all things begin and all things finish. It is so deeply embedded in human thinking that most people never question it. It is treated as a law of existence, as fundamental as gravity, as reliable as the sunrise.

But it is not a law. It is an observation. And observations made within a limited frame do not necessarily apply outside that frame.

Everything you have ever witnessed exists within a narrow band of scale, time, and energy. You live on one planet, orbiting one star, in one galaxy, during one tiny window of cosmic history. The things you observe beginning and ending are local phenomena. They are events happening within a system, not descriptions of the system itself. A wave begins and ends, but the ocean does not. A flame begins and ends, but energy does not. The question is whether the pattern you see at your scale holds at every scale, and there is no reason to assume it does.

Physics provides some interesting complications.

The first law of thermodynamics states that energy cannot be created or destroyed. It can change form. It can move from one system to another. But the total amount of energy in a closed system remains constant. If the universe is a closed system, then the energy within it has no beginning and no end. It has always existed in some form and will always exist in some form. It transforms, but it does not arrive or depart.

This directly contradicts the everyday intuition that everything begins and ends. The cup of coffee on your table began as water and ground beans and will end as a cold, stale liquid you pour down the sink. But the atoms in that coffee have existed for billions of years. The energy those atoms carry has existed for longer. The "beginning" and "end" of the coffee are descriptions of a temporary arrangement, not of the underlying substance. The arrangement starts and stops. The substance does not.

This distinction matters because it reveals that what we call a beginning is often just a reorganization, and what we call an end is just a dispersal. A star is born when gravity pulls hydrogen together and nuclear fusion ignites. A star dies when it runs out of fuel and collapses or explodes. But the hydrogen was there before the star, and the elements produced in the star's death persist long after. The star had a beginning and an end. The material did not.

So when we ask "does everything have a beginning and an end?" the answer depends entirely on what we mean by "everything" and what we mean by "beginning" and "end." If we mean specific arrangements of matter and energy, then yes, arrangements are temporary. If we mean the matter and energy themselves, then no, they appear to be permanent. The question is not one question. It is two questions wearing the same sentence.

The second law of thermodynamics adds another layer. It states that the total entropy of a closed system tends to increase over time. Entropy is often described as disorder, though that is a simplification. More precisely, entropy is a measure of the number of possible microscopic arrangements of a system. High entropy means many possible arrangements. Low entropy means few.

The universe began in an extraordinarily low entropy state. It has been increasing in entropy ever since. Stars burn out. Structures decay. Energy spreads out. The universe is slowly moving toward a state of maximum entropy, a condition sometimes called heat death, where no more work can be done, no more structures can form, and nothing changes. Not because everything has been destroyed, but because everything has reached equilibrium. Every temperature is the same. Every energy level is flat. The universe still exists, but nothing happens in it.

This is not the end of matter or energy. It is the end of activity. The end of difference. The end of events. The universe in heat death is still there. It still contains the same total energy it always did. But it is, for all practical purposes, dead. Not gone, but inert.

So does the universe have an end? In one sense, no. The matter and energy persist indefinitely. In another sense, yes. The capacity for anything to happen, for structures to form, for life to exist, for events to occur, that capacity has a finite lifespan. The universe will outlive its own ability to do anything. It will exist long after existence has any meaning.

This is a deeply uncomfortable idea, and it illustrates why the question "does everything have a beginning and an end?" is more complex than it appears. The universe does not end the way a candle ends. It does not go out. It fades. The distinction between existing and not existing becomes blurred. At what point do you say something has ended if the substance is still there but nothing is happening? Is a universe with no events still a universe?

There is no clear answer to that. It depends on how you define existence, how you define ending, and whether you consider persistence without activity to be a form of continuation or a form of death. These are not scientific questions. They are definitional ones. And definitional questions do not have correct answers. They have choices.

The same problem appears at the other end of the timeline. Did the universe have a beginning? The Big Bang model describes the earliest known state of the universe, but it does not describe a moment of creation from absolute non-existence. It describes a transition from one state to another. What came before that transition, if "before" even applies, remains unknown. Some models suggest the universe is cyclic, endlessly expanding and contracting. Others suggest it emerged from a prior quantum state. Others suggest time itself began with the Big Bang, making the concept of "before" meaningless.

None of these models confirm that the universe had a true beginning in the way we normally use the word. They describe boundaries of knowledge, not boundaries of existence.

The human fixation on beginnings and endings is not accidental. It is biological. Our brains are narrative engines. They process reality as a sequence of events with causes, developments, and conclusions. Every story has a beginning, a middle, and an end. Every memory is stored as a timeline. Every plan is structured as a progression from a starting point to a goal. The brain does not simply prefer narratives. It requires them. It cannot process reality without imposing a sequence on it.

This means that when we look at the universe, we automatically impose a narrative structure on it. We assume it must have started somewhere and must be heading somewhere. We assume it has a plot. We assume the plot has a resolution. And when we are told that it might not, that the universe might not have a beginning in any meaningful sense, or might not have an end in any final sense, the brain resists. Not because the evidence is unconvincing, but because the brain's operating system does not have a mode for processing stories without structure.

This is worth examining carefully, because it affects not only how we think about the universe but how we think about everything. Every culture in human history has created origin stories. Every religion has a creation account. Every civilization has a narrative about where we came from and where we are going. These are not failures of thinking. They are expressions of a deep cognitive need. The brain needs a beginning. It needs an end. It will generate them if they do not exist, because operating without them feels like falling.

But the universe is not obligated to conform to the structure of the human brain. The brain evolved to survive on the surface of a small planet, navigating social groups, avoiding predators, and finding food. Its narrative architecture is a survival tool, not a truth-detection instrument. The fact that the brain needs beginnings and endings does not mean reality provides them.

This creates a permanent tension between the mind and the cosmos. The mind demands a story. The cosmos does not tell one. The mind demands an origin. The cosmos may not have one. The mind demands a conclusion. The cosmos may simply continue, or fade, or transform, without ever arriving at anything the mind would recognize as an ending.

Mathematics, interestingly, has no problem with this. Mathematical structures can be infinite in both directions. The number line extends forever in both the positive and negative directions. There is no first number. There is no last number. No mathematician loses sleep over this. It is simply how the system works. Infinity is not a problem in mathematics. It is a feature.

But the human brain is not a mathematical system. It is a biological organ with finite processing power and a deep evolutionary investment in bounded narratives. Infinity in the abstract is manageable. Infinity applied to your own existence, to the question of whether everything you know and experience is a temporary arrangement in an endless process with no origin and no destination, that is something else entirely.

Some physicists have embraced this discomfort directly. Roger Penrose proposed the conformal cyclic cosmology model, in which the heat death of one universe becomes indistinguishable from the Big Bang of the next. In this framework, there is no true beginning and no true end. There are transitions. The end of one cycle looks, mathematically, identical to the start of another. The universe does not stop. It resets. Or more precisely, the distinction between stopping and starting dissolves at the appropriate scale.

This is an elegant model. It removes the need for a first moment and a last moment. But it does not resolve the underlying discomfort. It simply extends the timeline into infinity and asks you to accept that infinity is the answer. For many people, that is no more satisfying than being told the universe has a hard beginning and a hard end. The mind wants a boundary. Infinity refuses to provide one.

Religion, once more, offers a framework. In most religious traditions, there is a clear beginning (creation) and a clear end (judgment, apocalypse, dissolution, return to the divine). These narratives are psychologically powerful because they give the mind exactly what it craves: a structure, a purpose, a bounded story with a beginning, a middle, and an end. The universe was created by God and will end according to God's plan. Everything fits. Everything makes sense.

But when examined through the lens of the question we are asking, these narratives encounter the same wall. If God created the universe, then God existed before it. Did God have a beginning? If yes, the same question applies to God. If no, then something can exist without a beginning, which means the universe could also exist without a beginning, and God is not necessary for the explanation. If God will end the universe, what comes after? If something comes after, then the universe did not truly end. If nothing comes after, then the concept of ending leads back to the concept of nothing, which, as we explored in the previous chapter, may not be a coherent concept at all.

The religious framework provides a narrative. It provides comfort. It provides a bounded story. But it does not provide an answer that survives its own scrutiny. It closes the question by forbidding further inquiry, not by resolving it.

So where are we?

The first law of thermodynamics says energy has no beginning and no end. The second law says the universe's capacity for structure and activity is finite. Cosmological models disagree on whether the universe had a true origin or will have a true conclusion. The human brain demands both but has no means of verifying either. Religious frameworks provide both but cannot sustain them under examination. And the question itself turns out to be dependent on definitions that have no fixed meaning.

What counts as a beginning? The first appearance of matter? The first moment of time? The first event? What counts as an end? The last event? The cessation of all activity? The disappearance of matter itself? Each definition produces a different answer, and none of them is more correct than the others. They are choices, not discoveries.

The pattern by now should be familiar. A simple question, asked in plain language, containing hidden assumptions that do not hold. The assumption here is that "beginning" and "end" are objective, universal categories that apply to everything. They are not. They are human categories, shaped by human experience, enforced by human cognition, and projected onto a universe that may not operate in those terms.

Everything you have ever seen has had a beginning and an end. But everything you have ever seen is a tiny, local, temporary arrangement within a system that may have neither. Your experience is valid. Your generalization from it is not.

The question "does everything have a beginning and an end?" assumes the answer is yes or no. But the actual answer is: it depends on what you are asking about, what you mean by those words, and what scale you are examining. At the human scale, yes. At the cosmic scale, unclear. At the level of energy and fundamental law, probably not.

That is not a satisfying answer. It does not close cleanly. It does not provide the certainty the brain craves. It leaves the question partially open, leaning against a wall, unfinished.

Pointless, perhaps. But the question was never designed to be closed. It was designed to show you the gap between what the mind expects and what reality provides. And that gap is not a flaw. It is the territory this book lives in.

5

Where Did Matter Come From

Where did matter come from?

Look around you. Everything you can see and touch is made of matter. The walls, the floor, the air, your own body. Every object on Earth, every planet in the solar system, every star in every galaxy is composed of atoms, and those atoms are composed of smaller particles, quarks and electrons held together by fundamental forces. The sheer volume of matter in the observable universe is staggering. There are roughly ten to the power of eighty atoms in the visible universe. That is a one followed by eighty zeros. An incomprehensible quantity of physical substance, spread across billions of light years.

Where did all of it come from?

The scientific account of matter's history is remarkably detailed. We can trace it backward with impressive precision. The atoms in your body were assembled inside stars. Carbon, nitrogen, oxygen, iron, all of these were forged in the nuclear furnaces of massive stars over billions of years. When those stars died, either through supernova explosions or slower processes of shedding their outer layers, they scattered these elements into space. Gravity pulled the scattered material together again, forming new stars, new planets, and eventually the conditions necessary for life. You are, in a very literal sense, made of recycled star material.

Go further back. Before the first stars formed, the universe contained only the lightest elements: hydrogen, helium, and trace amounts of lithium. These were produced during Big Bang nucleosynthesis, a period lasting roughly three minutes that began about one second after the Big Bang. During this brief window, the universe was hot enough and dense enough for nuclear fusion to occur on a cosmic scale. Protons and neutrons combined into the nuclei of these light elements. After those three minutes, the universe expanded and cooled too much for further fusion. The heavier elements would have to wait for stars.

Go further still. Before nucleosynthesis, the universe was a soup of free protons, neutrons, electrons, and photons, all interacting in an extremely hot, dense plasma. Before that, in the first fractions of a second, conditions were so extreme that even protons and neutrons did not exist as stable particles. Instead, the universe was filled with a quark-gluon plasma, a state of matter in which the building blocks of protons and neutrons moved freely, not yet bound together.

And before that? In the earliest moments, during the first trillionth of a trillionth of a trillionth of a second, the universe underwent a period of rapid expansion called inflation. During this phase, the universe grew by a factor of at least ten to the power of twenty-six in an almost unimaginably short time. The energy driving this expansion is believed to have come from a quantum field called the inflaton field, though the precise nature of this field remains theoretical.

At the end of inflation, the energy stored in the inflaton field was converted into particles. This process, called reheating, is thought to be the moment when the matter and radiation content of the universe was produced. The energy of the field decayed into the fundamental particles that would eventually become everything we see today.

This is an extraordinary story. It spans from the first fraction of a second to the present day, across 13.8 billion years of cosmic evolution. It is supported by multiple independent lines of evidence: the cosmic microwave background radiation, the observed abundance of light elements, the large-scale structure of galaxies, and the redshift of distant objects. It is one of the great achievements of human knowledge.

And it does not answer the question.

It tells you how matter got from there to here. It tells you the sequence of transformations, the chain of events, the physical processes that turned energy into particles and particles into atoms and atoms into stars and stars into planets. It is a complete history of matter's journey through time.

But it is not a history of matter's origin. It is a history of matter's rearrangement.

The energy that was converted into particles during reheating had to come from somewhere. The inflaton field that drove inflation had to come from somewhere. The initial conditions of the universe, the precise configuration that allowed all of this to unfold, had to come from somewhere. And "somewhere" is exactly where the trail goes cold.

Physics can describe the transformation of energy into matter with extraordinary precision. Einstein's equation, energy equals mass times the speed of light squared, tells us that matter and energy are interchangeable. They are two forms of the same thing. Particle accelerators demonstrate this routinely: smash particles together with enough energy and new particles are created from the energy of the collision. Matter is not a separate substance from energy. It is energy in a particular configuration.

So the question "where did matter come from?" becomes "where did energy come from?" And that question, as we explored in the previous chapter, leads to the same dead end. If the total energy of the universe is zero, balanced between positive energy and negative gravitational energy, then in one sense, the energy did not come from anywhere. It adds up to nothing. But a zero composed of enormous positive and negative components is not the same as an undisturbed zero, and the question of why this particular arrangement exists rather than no arrangement at all remains unanswered.

Quantum field theory offers a partial framework. In this model, particles are not objects in the traditional sense. They are excitations of underlying fields. An electron is not a tiny ball floating in space. It is a ripple in the electron field. A photon is a ripple in the electromagnetic field. Matter is what happens when fields are disturbed in specific ways. The entire material universe is a pattern of disturbances in a collection of quantum fields.

This reframes the question. Instead of asking where matter came from, you could ask where the quantum fields came from. And instead of asking where the quantum fields came from, you could ask where the laws governing those fields came from. Each step pushes the question deeper, but no step answers it.

There is a specific problem within the story of matter's origin that deserves close attention, because it reveals just how much we do not understand.

In the standard model of particle physics, every particle has a corresponding antiparticle. An electron has a positron. A quark has an antiquark. A proton has an antiproton. When a particle and its antiparticle meet, they annihilate each other, converting their combined mass entirely into energy. And when energy converts into matter, it produces equal amounts of matter and antimatter. Always. Without exception. This is one of the most well-tested symmetries in physics.

This creates an enormous problem. If the Big Bang produced matter and antimatter in equal quantities, as the laws of physics say it should have, then every particle should have been annihilated by its antiparticle almost immediately. The universe should be filled with radiation and nothing else. No atoms. No stars. No planets. No people. Just light. An empty, glowing void with no structure and no substance.

But that is not what happened. Matter survived. Antimatter, for reasons that are not fully understood, did not. For every billion antimatter particles produced in the early universe, there were approximately one billion and one matter particles. That tiny asymmetry, one extra particle per billion, is the reason anything exists. Every atom in your body, every grain of sand, every galaxy in the sky is the leftover residue of that minuscule imbalance.

This is called the baryon asymmetry problem, and it is one of the great unsolved questions in physics. We know the asymmetry happened. We can calculate its magnitude. We can see its consequences everywhere we look. But we do not know why it happened. The standard model of particle physics does not fully account for it. There are proposed mechanisms, processes that could have produced a slight excess of matter over antimatter in the early universe, but none of them have been confirmed experimentally. They remain theoretical candidates, not established facts.

This means that the existence of all matter in the universe depends on a process we cannot explain. Not a process we have not yet observed. A process whose mechanism is fundamentally unknown. We know the result. We do not know the cause. And without that cause, the story of where matter came from has a gap in it that no amount of observational data has been able to fill.

It is worth pausing on this. The existence of matter, the existence of everything physical, every object and organism and structure in the cosmos, rests on a one-in-a-billion asymmetry whose origin is a mystery. If that asymmetry had not occurred, or had occurred in the opposite direction, the universe would contain nothing but radiation. The question "where did matter come from?" is not just a philosophical puzzle. It is a question about a specific physical event that determined the fate of reality, and we do not understand it.

Some physicists approach this through extensions of the standard model. Theories involving supersymmetry, leptogenesis, and various grand unified theories propose mechanisms that could generate the observed asymmetry. These are serious, mathematically rigorous proposals. They make predictions that could, in principle, be tested at particle accelerators or through cosmological observations. But so far, no experimental evidence has confirmed any of them. The Large Hadron Collider has not found supersymmetric particles. Leptogenesis remains a theoretical framework without direct observational support. The asymmetry remains unexplained.

Others look at it from a different angle entirely. If the universe is a simulation, then the asymmetry is simply a parameter. A setting. A value chosen by whoever or whatever designed the simulation. Matter exists because the simulation was configured to produce matter. The question of why becomes a question of design rather than physics. But as with every invocation of simulation theory, this does not answer the question. It transfers it. Why did the designers choose this configuration? What determines the parameters of the base reality in which the simulation runs? The regression continues.

Religion frames matter as a creation. God spoke the world into existence. God shaped clay into form. God willed matter into being. In these traditions, the question of where matter came from has a clear answer: it came from the will of a creator. But the will of a creator is not a physical mechanism. It is a narrative device. It tells you who is responsible without telling you how it was done. And the follow-up questions remain. Where did God get the material? If God created matter from nothing, then we return to the problem of something from nothing. If God created matter from divine substance, then where did that substance originate? The question does not close. It recedes.

There is also a more fundamental issue hiding beneath the surface of all these discussions. We tend to treat matter as the primary substance of reality. We ask where it came from because we assume it is the thing that needs explaining. But quantum field theory suggests that matter is secondary. Fields are primary. Particles are disturbances in fields. Matter is not the foundation of reality. It is a pattern on the surface of something deeper.

If that is correct, then asking where matter came from is like asking where the waves came from while ignoring the ocean. The waves are real. They have measurable properties. They behave in predictable ways. But they are not independent objects. They are expressions of a deeper system. And the real question is not about the waves. It is about the ocean. Where did the quantum fields come from? What determines their properties? Why do they exist at all?

Nobody knows. The fields are treated as fundamental in current physics. They are the starting point of the equations, not the result of them. They are not derived from something more basic. They are assumed. The laws that govern them are observed and described but not explained. Physics can tell you how the fields behave with extraordinary precision. It cannot tell you why they exist.

This is the wall again. The same wall that has appeared in every chapter so far. The chain of explanation is long, detailed, and supported by evidence. And at the end of it, the same silence. The trail goes cold not because of a lack of effort or imagination, but because there is a boundary beyond which the tools of human inquiry do not reach.

Where did matter come from? From energy. Where did energy come from? From quantum fields. Where did quantum fields come from? From the initial conditions of the universe. Where did those initial conditions come from?

The silence is the answer. Not because the answer is being withheld. Not because it exists somewhere waiting to be found. But because the question, pursued to its logical end, arrives at a point where explanation itself runs out of ground. The mind expects a foundation beneath the last foundation. Reality does not appear to provide one.

Every atom in your hand has a traceable history stretching back billions of years. Every element was forged in a specific place at a specific time through a specific process. The story is complete in every detail except the first one. The opening line is missing. The first page is blank.

And no matter how many chapters we add to the middle and the end, that blank page remains. Staring back. Pointless to ask about, perhaps. But impossible to stop asking.

6

What Is "Nothing"

What is nothing?

Not what happens when there is nothing. Not what existed before something. Not the absence of a particular object or substance. The question is more direct than that. What is nothing itself? Can you define it? Can you describe it? Can you think about it without turning it into something?

This is not a trick question. It is a genuine problem that has occupied philosophers for over two thousand years and physicists for the better part of a century. It sounds like it should be the simplest concept imaginable. Nothing is the absence of everything. Empty. Void. Blank. Zero. The word is used constantly in everyday language without any difficulty. There is nothing in the fridge. Nothing happened today. I have nothing to say.

But in each of these cases, "nothing" does not actually mean nothing. It means the absence of something specific. There is no food in the fridge, but there is air, light, shelves, cold. Nothing happened today means nothing noteworthy happened, but you still breathed, your heart still beat, the planet still turned. I have nothing to say means I have no words I want to share, not that language and thought have ceased to exist.

Every common use of the word "nothing" is a relative nothing. An absence measured against an expectation. True nothing, absolute nothing, would mean the absence of everything. No matter. No energy. No space. No time. No dimensions. No laws. No potential. No properties of any kind. Not even the possibility of properties. Not even the concept of absence.

Try to picture that.

You cannot. The moment you try, the mind generates a space. A dark, empty room. A void. A black expanse stretching in all directions. But that image is not nothing. It is a space without objects. It still has dimensions. It still has extent. It still has darkness, which is a visual property. The mind cannot picture true nothing because the act of picturing requires a frame, and a frame is something.

This is not a failure of imagination. It is a structural limitation of consciousness. Consciousness is always conscious of something. It operates by representing objects, spaces, ideas, and sensations. It cannot represent the total absence of all representation. Asking the mind to conceive of true nothing is like asking an eye to see the absence of all light, not darkness, because darkness is a visual experience, but the complete absence of vision itself. The tool cannot perform the task because the task requires the tool to negate itself.

Philosophy has circled this problem endlessly. Parmenides, as mentioned in an earlier chapter, argued that nothing cannot exist because to speak of it or think of it is to make it a subject, and a subject is something. His argument has never been fully refuted. It has been restated, reformulated, and challenged from various angles, but the core problem remains. Nothing, the moment it is discussed, becomes a concept. And a concept is not nothing.

Some philosophers have tried to resolve this by distinguishing between nothing as a concept and nothing as a state. The concept of nothing exists, they argue, but the state of nothing does not. You can talk about nothing the way you can talk about unicorns. The word refers to something in the realm of ideas, but not to anything in the realm of reality. Nothing is a useful abstraction, a placeholder, a way of marking the absence of something specific. But it does not correspond to an actual condition that has ever existed or could ever exist.

This is a reasonable position, but it has implications that most people do not follow to their conclusion. If true nothing cannot exist, then something has always existed. There was never a state of absolute absence. There was never a time when there was no time. There was never a condition with no conditions. Reality, in some form, is permanent. It does not have a starting point because there was never an absence for it to start from.

This brings us back to the question from Chapter 3, but from the other side. That chapter asked whether something can come from nothing. This chapter asks whether nothing is even a coherent concept. If it is not, then the question of something coming from nothing was never valid to begin with. It was a question about a state that cannot exist, posed in language that made it sound real.

Physics approaches nothing differently, and the results are equally unsettling.

In classical physics, a vacuum is the closest approximation to nothing. A region of space with all matter and energy removed. Empty space. But even in classical physics, the vacuum is not truly empty. It still has spatial dimensions. It still has time. It still has the properties of geometry and distance. It is a container with nothing in it, but the container itself is something.

Quantum mechanics makes this far more complicated. In quantum field theory, the vacuum state is the lowest energy state of a quantum field. It is not empty. It is active. It seethes with virtual particles, temporary fluctuations that arise from the uncertainty principle. These fluctuations are not hypothetical. Their effects are measurable. The Casimir effect, in which two uncharged metal plates placed very close together in a vacuum are pushed toward each other by the pressure of virtual particles, has been confirmed experimentally. The vacuum is not nothing. It is a dynamic, energetic, fluctuating state with real physical consequences.

This means that the closest thing to nothing that physics can describe is still something. It has structure. It has properties. It has behavior. It obeys laws. It is a specific state of a specific system, and it is as far from true nothing as a quiet ocean is from the absence of water.

Lawrence Krauss famously argued that the universe could have come from nothing, where "nothing" was defined as the quantum vacuum. His argument was met with immediate criticism, not from the fringes of science, but from physicists and philosophers who pointed out that calling a quantum vacuum "nothing" is a misuse of the word. A quantum vacuum has properties. It has energy. It has fluctuations. It obeys the laws of quantum field theory. It exists within a mathematical framework. It is, by any ordinary definition of the word, something.

Krauss responded that this is the only kind of nothing that physics can address. And he was right. Physics cannot describe true nothing because true nothing has no variables, no equations, no measurable quantities. It has nothing for physics to work with. The discipline was built to describe reality, and nothing is the absence of reality. Asking physics to define nothing is asking a net to catch the absence of fish.

This is not a criticism of physics. It is a recognition of its boundaries. Every tool has a domain, and the domain of physics is the existing universe. What lies outside that domain, if anything, is not a question physics can answer. It can describe the vacuum. It can describe the quantum fields. It can describe the lowest possible energy state of the cosmos. But it cannot describe the total absence of all of these things, because that absence, if it is even possible, has no physical description.

Philosophy pushes further, but does not fare much better.

Existentialist thinkers, particularly Sartre, drew a distinction between being and nothingness. For Sartre, nothingness was not a state of the world but a feature of consciousness. The ability to conceive of absence, to imagine what is not there, to negate what is present, this capacity is what separates human consciousness from mere existence. A rock does not experience nothing. A rock simply is. A human being can look at a room and notice what is missing. That act of noticing, of recognizing absence, is the origin of nothingness in Sartre's framework.

This is a psychological account, not a cosmological one. It tells you about the mind's relationship with absence, not about whether absolute absence is a real state of affairs. But it contains an insight that is relevant to this chapter. The concept of nothing may be entirely a product of consciousness. It may be something the mind generates, not something the mind discovers. There may be no nothing out there in reality. There may only be the mind's ability to subtract things from its model of the world and label the result as nothing.

If that is the case, then asking "what is nothing?" is not asking about reality. It is asking about a mental operation. It is asking about subtraction, not about a destination that subtraction leads to. And that mental operation, no matter how far you extend it, never actually arrives at true nothing. You can subtract objects from a room. You can subtract air. You can subtract light. You can subtract space. You can subtract dimensions. But at every stage, you are still performing the subtraction, and the act of performing it requires a mind, which is something. The process never completes. It approaches nothing asymptotically, getting closer and closer but never arriving.

This is not just a philosophical curiosity. It has direct implications for the biggest questions in cosmology. If true nothing is incoherent, then the universe did not come from nothing. Not because it came from something specific, but because the concept of a state from which it could have emerged, defined as the total absence of everything, does not function. The origin question does not have a hidden answer. It has a broken premise.

Some theologians have used this to argue for the existence of God. If nothing cannot exist, then something must be eternal. And that eternal something, they argue, is God. This is structurally identical to the cosmological argument we have already examined. It works as a statement of logical necessity, but it does not identify what the eternal something is. Calling it God gives it a name without giving it an explanation. And the question of what God is, what God is made of, where God exists, how God functions, leads to the same recursive problems that have appeared in every chapter.

There is also a mathematical dimension to this question that is rarely discussed in popular treatments. In set theory, the empty set is a well-defined mathematical object. It is the set that contains no elements. It exists within mathematics as a valid, useful concept. From the empty set, mathematicians can construct the natural numbers, and from the natural numbers, they can build the entire edifice of mathematics. In a sense, all of mathematics emerges from nothing, from the empty set, from the formalization of absence.

But the empty set is not nothing. It is a mathematical object with properties. It has a cardinality of zero. It is a subset of every set. It is unique. It is defined by what it does not contain, but the definition itself is a form of content. The empty set is the mathematical version of the same paradox we have been examining. It is the closest formal representation of nothing, and it is still something.

Every discipline that attempts to define nothing ends up in the same place. Physics defines the vacuum, which is something. Philosophy defines the concept of absence, which is something. Mathematics defines the empty set, which is something. Theology defines the state before creation, which requires a creator, which is something. Every attempt to reach nothing stops short, because the tools used to reach it are themselves things, and they cannot negate themselves out of existence.

This suggests a possibility that is difficult to accept but may be correct. Nothing might not be a real state. It might not be a place the universe came from or a condition that could ever obtain. It might be purely a linguistic construction, a word that functions in everyday speech but collapses when applied to fundamental questions about reality. Like a mirage that looks solid from a distance but dissolves when you walk toward it.

If nothing is not real, then every question that assumes it is, every question about what existed before the universe, what the universe emerged from, what would remain if everything were removed, is built on a false foundation. Not wrong in its intention. Not foolish in its motivation. But structurally incapable of producing an answer, because the thing it asks about does not exist in the way the question assumes.

What is nothing?

It might be the one thing that truly has no answer. Not because the answer is hidden. Not because we lack the tools to find it. But because the question, when examined closely, turns out to be asking about something that cannot exist, described by a word that only functions when you do not look at it too carefully. A concept that dissolves under its own weight.

A pointless question, then. Unless the point was never to answer it, but to discover what happens when you try.

7

Why Do We Need Answers

Why do we need answers?

Every chapter so far has ended in the same place. A question is asked. Attempts are made. The attempts fail. The question remains open. And yet, despite knowing this pattern by now, despite recognizing that the next chapter will do the same thing, you are still reading. Something in you has not accepted the situation. Something in you is still waiting for the moment when the pieces fall into place, when the fog clears, when the answer finally arrives.

That impulse is worth examining, because it may be the most important question in this book. Not what was before the universe. Not what nothing is. Not where matter came from. But why, when faced with questions that cannot be answered, the human mind refuses to stop trying.

This is not a philosophical abstraction. It is a biological fact. The human brain is an answer-seeking machine. It was built for it. Millions of years of evolution have shaped a cognitive system that is relentlessly, compulsively oriented toward finding solutions. Identifying patterns. Resolving uncertainty. Closing open loops. This is not a preference. It is not a cultural habit. It is wired into the architecture of the organ itself.

The neurological basis for this is well documented. When the brain encounters an unanswered question or an unresolved problem, it enters a state of cognitive tension. This is not a metaphor. It is a measurable change in brain activity. The prefrontal cortex becomes more active. Dopamine pathways are engaged. The brain begins allocating resources toward solving the problem, the same way it would allocate resources toward finding food or avoiding a threat. Uncertainty is treated, neurologically, as a form of danger. The brain does not distinguish between a physical threat it cannot identify and an intellectual question it cannot answer. Both produce the same underlying response: find the answer, resolve the tension, restore equilibrium.

This mechanism is extraordinarily useful in the environment it evolved for. On the savannah, uncertainty could kill you. A rustling in the grass could be the wind or it could be a predator. The brain that resolved that uncertainty fastest survived. The brain that tolerated ambiguity died. Over millions of years, this pressure produced a cognitive system that is deeply, fundamentally intolerant of unanswered questions. Not because unanswered questions are inherently dangerous in the modern world, but because the brain still runs on software designed for a world where they were.

This is why unsolved problems feel uncomfortable. Not intellectually uncomfortable, but physically uncomfortable. An unresolved question produces a low-grade stress response. It nags. It lingers. It occupies processing power even when you are not consciously thinking about it. Psychologists call this the Zeigarnik effect: the tendency for incomplete tasks to remain more active in memory than completed ones. A waiter remembers unfilled orders better than filled ones. A student remembers unanswered exam questions better than answered ones. The brain prioritizes the unfinished because the unfinished represents a potential threat that has not been neutralized.

This means that when you read a chapter of this book and reach the end without an answer, your brain does not simply note the absence and move on. It flags it. It stores it. It keeps a thread running in the background, searching for resolution. And when the next chapter begins with a new question, the brain accepts it because it believes this time, maybe, the answer will come. This time, the loop will close.

It will not. But the brain cannot accept that in advance. It has to try.

This is not a flaw. In most contexts, it is a strength. The refusal to accept the absence of an answer is what drives scientific discovery, engineering innovation, medical research, and every other form of progress. The polio vaccine exists because someone refused to accept that polio was unanswerable. The airplane exists because someone refused to accept that human flight was impossible. Every breakthrough in human history began with someone looking at an unsolved problem and refusing to leave it alone.

But the same mechanism that produces breakthroughs also produces something else: false answers. Because the brain cannot tolerate unanswered questions, it will generate answers even when no real answer exists. It will fill the gap with something, anything, rather than leave it empty. A pattern that is not there. A cause that does not exist. A story that feels true but is not.

This is the origin of superstition, conspiracy theory, and a significant portion of religious belief. Not all of it. There are serious, rigorous theological and philosophical traditions that engage with these questions honestly. But a large proportion of the answers that human beings have generated for unanswerable questions are not the product of careful thought. They are the product of a brain that cannot tolerate the empty space and fills it with whatever material is available.

Thunder is frightening and unexplained, so a god must be throwing it. Disease strikes without warning, so someone must have cursed the village. The harvest fails, so the spirits must be angry. These are not stupid conclusions. They are the natural output of a cognitive system that demands causes for effects and will manufacture them when they are not apparent.

Modern versions of the same process are more sophisticated but structurally identical. We do not know what preceded the Big Bang, so perhaps the universe is a simulation. We do not know why consciousness exists, so perhaps it is a quantum phenomenon. We do not know whether the multiverse is real, so perhaps every possibility is actualized somewhere. Each of these is a legitimate hypothesis explored by serious thinkers. But each of them also functions, psychologically, as a gap-filler. They give the mind something to hold onto in the absence of solid ground.

The question is whether that is a good thing or a bad thing. And the honest answer is: both.

The drive to find answers has produced everything valuable in human civilization. Every technology, every medicine, every scientific law, every mathematical theorem began as a question that someone refused to leave unanswered. Without that drive, we would still be sheltering in caves, interpreting every natural event as the whim of spirits. The compulsion to resolve uncertainty is the engine of progress. It is arguably the single most important feature of the human mind.

But it has a cost. And the cost is that the mind cannot tell the difference between a question that has not yet been answered and a question that cannot be answered. It treats both the same way. It applies the same urgency, the same cognitive resources, the same emotional pressure to both. A question that will be answered in ten years and a question that will never be answered in any number of years feel identical from the inside. The brain has no mechanism for distinguishing them.

This matters because the two categories require completely different responses. A question that has not yet been answered requires persistence. Keep working. Keep investigating. The answer is out there and will eventually be found. A question that cannot be answered requires acceptance. Stop working. Recognize the boundary. Redirect your resources elsewhere.

But the brain does not accept boundaries. It was not built for acceptance. It was built for persistence. And so, when faced with a question that genuinely cannot be answered, it does what it always does: it keeps trying. It cycles through possible solutions. It generates hypotheses. It constructs frameworks. It consults authorities. It argues. It debates. It lies awake at night turning the problem over. And none of it produces an answer, because no answer exists. But the brain cannot stop, because stopping feels like failure, and failure feels like danger, and danger activates the very system that is already running at full speed.

This is the loop that traps people when they engage seriously with the questions in this book. What was before the universe? The mind tries to answer it. Fails. Tries again. Fails again. Tries a different angle. Fails from that angle too. The discomfort builds. The mind reaches for something, anything, that will make the discomfort stop. And whatever it grabs, whether it is a religious explanation, a scientific hypothesis, or a philosophical framework, it holds onto with a grip that has less to do with the quality of the explanation and more to do with the intensity of the need.

This is why people argue so passionately about questions that have no answers. It is not because they are certain they are right. It is because letting go of their answer would mean returning to the discomfort of not having one. The answer is not held because it is convincing. It is held because the alternative is unbearable.

This applies to scientists as much as it applies to anyone else. Scientists are human beings with human brains running the same cognitive software as everyone else. A physicist who has spent thirty years working on string theory has a neurological investment in string theory being correct that extends far beyond the mathematics. Abandoning it would mean reopening questions that the theory had closed, and the brain resists that reopening with the same force it uses to resist any return to uncertainty. This is not a criticism of scientists. It is a description of how human cognition works, regardless of training or discipline.

The philosopher Karl Popper recognized this problem in the structure of science itself. He argued that a theory is only scientific if it can, in principle, be falsified. If no possible observation could prove it wrong, it is not a scientific theory. It is a belief system wearing the language of science. This criterion was designed to protect science from the brain's tendency to cling to answers that feel right but cannot be tested. It is a guardrail against the very impulse we are discussing.

But guardrails only work if people stay behind them. And in practice, the boundary between testable theory and untestable speculation is not always clear. String theory makes predictions, but those predictions require energy levels far beyond what any current or foreseeable technology can reach. The multiverse is a logical consequence of certain inflationary models, but no observation from within our universe can confirm or deny the existence of other universes. Simulation theory is consistent with certain features of quantum mechanics, but no test can determine from inside a simulation whether the simulation is real.

These are not fringe ideas. They are discussed in mainstream physics and philosophy. But they share a common feature: they cannot currently be tested, and it is not clear they ever will be. They exist in a space between science and speculation, between hypothesis and belief. And the reason they persist in that space, the reason serious thinkers continue to develop and defend them, is not purely rational. It is partly neurological. The brain needs answers. These provide answers. The fact that the answers cannot be verified is, for the brain, a secondary concern.

This is the uncomfortable truth at the center of this chapter. The need for answers is not a path to truth. It is a path to relief. Sometimes truth and relief coincide. Sometimes the answer that makes the discomfort stop is also the answer that is correct. But not always. And in the case of the questions this book is asking, possibly not ever.

The questions in this book are not puzzles waiting to be solved. They are boundaries. They mark the edge of what human cognition can process. And the appropriate response to a boundary is not to keep running into it. It is to acknowledge it, examine it, understand why it is there, and then decide what to do with that knowledge.

But the brain will not do that voluntarily. It has to be shown. It has to be walked up to the edge repeatedly, allowed to try, allowed to fail, and then shown that the failure is not a deficiency. It is a feature of the landscape. The wall is not there because you are not smart enough to get past it. The wall is there because there is nothing on the other side. Not nothing as in empty space. Nothing as in the absence of a place to go.

This is what frustration feels like when it has nowhere to land. Not the frustration of a problem that is too hard. The frustration of a problem that is not a problem at all, but a boundary disguised as a question. The mind reaches for the answer and finds that there is no shelf to place it on. No hook to hang it from. The architecture for receiving the answer does not exist.

Why do we need answers? Because the brain was built to need them. Because evolution made uncertainty feel like danger. Because the cognitive machinery that kept our ancestors alive on the savannah does not have an off switch for abstract questions about the nature of reality.

But needing an answer and there being an answer are two entirely different things. The need is real. It is biological, measurable, and powerful. The answer is not guaranteed. The universe is not obligated to provide one just because your brain demands it.

And sitting with that, truly sitting with it, not rushing past it to the next hypothesis, not reaching for a comforting framework, not arguing your way out of the discomfort, just sitting with the raw fact that some questions will never be answered and your brain will never stop wanting them to be, that is perhaps the most difficult thing this book will ask you to do.

Not because it is intellectually complex. But because every cell in your brain is screaming at you to keep looking.

8

The Illusion of Understanding

You understand gravity. Or at least, you think you do.

An apple falls from a tree. You catch it or you do not. Either way, you understand what happened. The apple was pulled downward by gravity. It accelerated toward the ground at roughly 9.8 meters per second squared. You learned this in school. You have experienced it every day of your life. You have never floated off the surface of the Earth. Objects fall. Gravity pulls. You understand it.

Except you do not.

You understand what gravity does. You do not understand what gravity is. You can describe its effects with perfect accuracy. You can calculate trajectories, predict orbits, and land a spacecraft on a comet millions of kilometers away using gravitational equations. But if someone asks you what gravity actually is, what it is made of, why it exists, how it works at the most fundamental level, the honest answer is that nobody knows.

Newton described gravity as a force acting at a distance between two masses. His equations work brilliantly for most practical purposes. They got us to the moon. But Newton himself admitted that he did not know how gravity operated. He could describe the relationship between mass, distance, and gravitational force with mathematical precision. He could not explain the mechanism. How does one object "know" another object is there? How does the force travel across empty space instantaneously? Newton had no answer. He called it a question he would leave to the reader's consideration. That was in 1687. The question has not been answered since.

Einstein improved the description. In general relativity, gravity is not a force in the traditional sense. It is a curvature of spacetime caused by the presence of mass and energy. Objects do not fall because they are pulled. They fall because they follow the curved geometry of the space around them. The Earth does not pull the apple. The Earth warps the fabric of spacetime, and the apple follows the curve.

This is a more accurate description than Newton's. It explains phenomena that Newton's model could not, such as the bending of light around massive objects and the precise orbit of Mercury. It has been confirmed by experiment after experiment. It is, by every scientific standard, a better model.

But it is still a description. It tells you how gravity behaves. It tells you the shape of the effect. It does not tell you why mass curves spacetime. It does not tell you what spacetime is made of. It does not explain the mechanism by which the presence of matter alters the geometry of the universe. General relativity is a map of extraordinary accuracy. But a map is not the territory. And the territory, in this case, remains unexplored.

This is what the illusion of understanding looks like. You can predict every effect of gravity with near-perfect precision. You can build bridges, fly aircraft, and send satellites into orbit. Your practical mastery is complete. And that practical mastery creates the feeling that you understand the thing itself. But you do not. You understand its behavior. You have mistaken the pattern for the explanation.

This illusion extends far beyond gravity. It covers almost everything.

You understand electricity. You flip a switch, the light comes on. Electrons flow through a circuit. The mechanism is clear. But what is an electron? It is a fundamental particle with a specific charge and mass. What is charge? It is a property of certain particles that determines their electromagnetic interactions. Why do some particles have charge and others do not? That is not explained. It is observed. What is mass? In the standard model, mass is generated through interaction with the Higgs field. What is the Higgs field? It is a quantum field that permeates all of space. Why does it exist? Unknown.

Every explanation, followed far enough, terminates in an observation. Not in an understanding. The chain of "why" questions always ends at the same place: this is how it is, and we do not know why. The final link in every explanatory chain is not an answer. It is a description of something we have observed but cannot explain.

Physics is not unique in this. It is simply the discipline where the problem is most visible because physics deals with fundamental questions and has the most rigorous standards of evidence. But the illusion of understanding operates everywhere.

You understand evolution. Organisms with traits better suited to their environment survive and reproduce at higher rates, passing those traits to their offspring. Over time, this process produces the diversity of life. Clear. Simple. Well-supported by evidence. But why do mutations occur? Because of errors in DNA replication. Why do errors in DNA replication occur? Because the molecular machinery is not perfect. Why is it not perfect? Because perfection would require a level of energy expenditure that is not sustainable. Why? Because of the thermodynamic properties of the system. Why does the system have those properties? Because of the fundamental constants of physics. Why do those constants have those values?

The chain ends. Again. Always in the same place. A description of what is, with no explanation of why.

You understand consciousness. Or rather, you experience consciousness, and you assume that experience constitutes understanding. You are aware. You have thoughts, feelings, perceptions. You know what it is like to be you. But what is consciousness? How does subjective experience arise from physical processes? How does a collection of neurons firing electrical signals produce the experience of seeing red, tasting salt, or feeling grief?

This is known as the hard problem of consciousness, and it remains entirely unsolved. Not partially solved. Not nearly solved. Entirely unsolved. There is no accepted theory of how physical processes give rise to subjective experience. There are hypotheses, frameworks, and models. Some are more popular than others. None of them have been confirmed. The gap between the physical description of brain activity and the subjective experience of being alive has not been bridged. It has not even been narrowed in any measurable way.

And yet most people feel they understand consciousness because they have it. The experience of being conscious creates the illusion of understanding it. This is like saying you understand how a television works because you can watch the programs. The experience of the output tells you nothing about the mechanism that produces it.

The illusion of understanding is not a minor cognitive quirk. It is one of the most pervasive and consequential features of human thinking. It shapes how we engage with science, religion, politics, and each other. It determines what questions we ask and, more importantly, what questions we never think to ask because we believe they have already been answered.

Psychologists have studied this extensively. The phenomenon is sometimes called the illusion of explanatory depth. In experiments, people are asked to rate how well they understand common objects or processes, things like zippers, toilets, bicycles, or the tidal system. Most people rate their understanding quite highly. Then they are asked to explain, in detail, how the thing actually works. Step by step. Mechanism by mechanism. Almost without exception, their explanations collapse within the first few sentences. The understanding they thought they had evaporates the moment they are required to articulate it.

This is not because people are ignorant. It is because the brain confuses familiarity with comprehension. You have seen a toilet flush a thousand times. You know what happens when you press the handle. Water moves. Things disappear. The bowl refills. You have watched the sequence so many times that it feels understood. But watching something happen and understanding why it happens are entirely different cognitive acts. The brain blurs the boundary between them because, for most practical purposes, the distinction does not matter. You do not need to understand fluid dynamics to use a toilet. You just need to know which handle to push.

This works fine for everyday life. It does not work for fundamental questions about reality.

When someone says they understand the Big Bang, what they usually mean is that they have heard a description of it and the description made sense to them at the time. The universe was hot and dense and then it expanded. Simple enough. But the feeling of understanding that description produces is not the same as actually understanding what happened. What does it mean for all of space to be compressed into a point? What physical mechanism caused the expansion? What determined the laws that governed the process? The description answers none of these questions. It provides a narrative, and the brain mistakes the narrative for comprehension.

Religion operates on the same principle, often more effectively. Religious explanations of the universe are typically simpler, more emotionally resonant, and more narratively satisfying than scientific ones. God created the world. God has a plan. Everything happens for a reason. These are extraordinarily powerful narratives. They provide a complete framework: an origin, a purpose, a destination. The brain receives them and feels the relief of understanding. The loop closes. The tension dissolves. The question feels answered.

But the feeling of an answer is not an answer. It is a neurological event. It is the brain's reward system responding to the closure of an open loop. It feels exactly the same whether the answer is correct, incorrect, or meaningless. The sensation of understanding is not evidence of understanding. It is evidence that the brain has found something it is willing to accept.

This is why people can hold completely contradictory beliefs with equal confidence. A young-earth creationist feels they understand the origin of the universe. A theoretical physicist feels they understand the origin of the universe. Both experience the same subjective sensation of comprehension. Both would rate their understanding highly. And both, if pressed far enough on the details, would eventually reach a point where their explanation stops and they have nothing left but an assertion. God willed it. The quantum fields fluctuated. Neither can go further. But both feel like they can, because the illusion of understanding fills the gap between where their explanation ends and where they believe it ends.

This illusion is self-reinforcing. Once the brain accepts an explanation, it filters subsequent information through that explanation. Confirming evidence is noticed and remembered. Contradicting evidence is minimized or dismissed. This is confirmation bias, and it works in tandem with the illusion of understanding to create a sealed system. You believe you understand something. You notice evidence that supports your understanding. You dismiss evidence that contradicts it. Your belief in your own understanding strengthens. And with each cycle, the illusion becomes harder to penetrate.

Breaking through this requires a specific and uncomfortable act: admitting that you do not understand something you thought you did. This is cognitively expensive. It means reopening a closed loop. It means reintroducing uncertainty into a space where the brain had achieved equilibrium. It means going backward, from the comfort of feeling like you know something to the discomfort of recognizing that you do not. The brain resists this powerfully. It does not want to reopen settled questions. It does not want to return to uncertainty. It would rather maintain a wrong answer than face the discomfort of no answer.

This is the mechanism that makes this book difficult to read. Not the vocabulary. Not the concepts. The concepts are straightforward. What is difficult is that each chapter takes something the reader feels they understand, opens it up, shows them the emptiness inside, and then refuses to close it again. The brain wants closure. The book denies it. That denial produces discomfort, and that discomfort is the point.

Because the illusion of understanding is not harmless. It does not only affect individuals. It shapes entire cultures. It determines which questions get funded, which research gets pursued, which ideas are taken seriously, and which are dismissed. When a society collectively believes it understands something, it stops investigating it. The question is closed. Resources move elsewhere. And if the understanding was an illusion, the question remains unanswered but is now also ignored.

For centuries, the mechanics of disease were "understood" as imbalances in bodily humors. This illusion prevented the development of germ theory. For decades, the structure of the atom was "understood" as a miniature solar system with electrons orbiting a nucleus. This illusion obscured the reality of quantum behavior. For most of human history, the Earth's position at the center of the universe was "understood" as obvious and self-evident. Each of these illusions felt like knowledge. Each of them was a wall painted to look like a window.

The question this chapter poses is not whether you understand the specific topics discussed in this book. It is whether you understand anything as deeply as you believe you do. The answer, almost certainly, is no. Not because you are unintelligent. But because the brain is designed to make you feel like you understand things, regardless of whether you actually do. That feeling is a tool. It allows you to function in the world without being paralyzed by the complexity of everything around you. But it is not truth. It is convenience.

And the distance between convenience and truth, between feeling like you understand and actually understanding, is where most of the unanswered questions in this book live. They sit in that gap, invisible to anyone who has not yet realized the gap exists, painfully obvious to anyone who has.

The illusion of understanding is the most comfortable lie the brain tells itself. And it tells it constantly, effortlessly, and without any awareness that it is doing so.

Recognizing that lie is not the same as fixing it. You cannot rewire millions of years of evolution by reading a chapter in a book. The illusion will return the moment you close these pages. It will reassert itself the next time someone asks you a question and your brain offers a confident answer before you have had time to examine whether the confidence is earned.

But for a moment, right now, sitting with the possibility that most of what you think you understand is a description mistaken for an explanation, you are closer to an honest relationship with reality than you usually are.

That is not an answer. It is not even progress. But it is something most people never experience, because the illusion is too comfortable to question and too effective to notice.

9

The Limits of Science

Science is the most successful method of understanding reality that human beings have ever developed. Nothing else comes close. In roughly four centuries of modern scientific practice, humanity has gone from not knowing what stars are to mapping the cosmic microwave background radiation left over from the first moments of the universe. We have decoded the structure of DNA. We have split the atom. We have eradicated diseases that killed millions. We have sent machines beyond the edge of the solar system. The track record is extraordinary, and it earns science a level of trust that no other method of inquiry can claim.

But trust and limitlessness are not the same thing.

Science has boundaries. Not temporary boundaries that will be pushed back with more funding and better technology. Structural boundaries. Boundaries that exist because of what science is, how it works, and what it requires in order to function. These boundaries are rarely discussed in popular science, because acknowledging them feels like an admission of weakness, and in a culture that often treats science as the final authority on all questions, weakness is not welcome.

But ignoring the limits of science does not make them disappear. It just makes them invisible. And invisible limits are the most dangerous kind, because you do not know when you have crossed them.

The first and most fundamental limit of science is that it can only address questions that are empirically testable. A question is scientific if, and only if, it is possible to design an observation or experiment that could, in principle, confirm or refute the answer. If no such observation is possible, the question falls outside the domain of science. Not because the question is unimportant. Not because it is meaningless. But because science has no tool for engaging with it.

This seems like a minor restriction until you consider how many important questions fail the test.

Is there a purpose to the universe? Science cannot address this. Purpose is not measurable. It does not produce a signal that an instrument can detect. You cannot design an experiment to test whether the cosmos has an intention behind it. The question is real. People care about it deeply. It affects how they live their lives, raise their children, and face their deaths. But science has nothing to say about it. Not because the answer is no. Because science cannot reach the question.

Is consciousness fundamental to reality, or is it a byproduct of physical processes? Science can study the neural correlates of consciousness. It can identify which brain regions are active during specific experiences. It can map the relationship between physical states and reported subjective states. But it cannot access subjective experience directly. It cannot measure what it feels like to see blue. It cannot quantify grief. It cannot test whether a sufficiently complex computer would experience anything at all. The hard problem of consciousness sits outside the reach of any current scientific method, and it is not clear that any future method could reach it either, because the problem is not about gathering more data. It is about a category of phenomenon that may not be reducible to data.

Do other universes exist? Several well-regarded physical theories predict or allow for the existence of other universes. Eternal inflation suggests an infinite landscape of bubble universes. The many-worlds interpretation of quantum mechanics posits that every quantum event splits reality into branches. String theory allows for an enormous number of possible vacuum states, each potentially corresponding to a different universe. These are not speculations from the margins of science. They come from the core of theoretical physics.

But none of them can be tested. No signal from another universe has ever been detected. No experiment can access a parallel branch of quantum reality. No observation can confirm or deny the existence of a universe beyond our own. These theories make predictions that are consistent with what we observe, but consistency is not confirmation. A theory that cannot be distinguished from its alternatives by any possible observation is not, by Popper's criterion, a scientific theory. It is something else. It may be true. It may be useful. It may be mathematically beautiful. But it is not science in the strict sense of the word.

This does not mean these ideas should be abandoned. It means they should be recognized for what they are: theoretical frameworks that have crossed the boundary between empirical science and mathematical philosophy. They operate in a space where the normal rules of scientific validation do not apply, because the normal rules require the possibility of being wrong, and these theories cannot be shown to be wrong by any currently conceivable means.

The second major limit of science is the problem of induction. Science works by observing patterns and generalizing from them. You drop a ball a thousand times. It falls every time. You conclude that balls fall when dropped. This seems reasonable, and it works in practice. But philosophically, it is not secure. The fact that something has happened a thousand times does not guarantee it will happen the thousand-and-first time. The sun has risen every morning for the entirety of human history. This does not prove, in the strict logical sense, that it will rise tomorrow. It makes it overwhelmingly likely. But likelihood is not certainty, and science, despite its enormous predictive power, operates on likelihood, not certainty.

David Hume identified this problem in the eighteenth century, and it has never been resolved. Science assumes that the laws of nature are consistent across time and space. It assumes that an experiment performed tomorrow will yield the same result as the same experiment performed today. It assumes that the physical constants measured in this galaxy hold in distant galaxies. These assumptions are supported by an overwhelming body of evidence. But they are still assumptions. They cannot be proven from first principles. They are articles of faith, dressed in the language of methodology.

This is not a popular thing to say. The word "faith" applied to science makes many scientists uncomfortable, because faith is associated with religion, and science defines itself in opposition to faith. But the discomfort does not change the reality. Science rests on foundational assumptions that cannot be verified by the scientific method itself. The uniformity of nature. The reliability of observation. The validity of mathematical description. These are the bedrock on which everything else is built, and they are accepted, not proven.

The third major limit of science is perhaps the most unsettling, because it does not come from philosophy or methodology. It comes from physics itself. And it has a name: quantum mechanics.

Quantum mechanics is the most successful predictive theory in the history of science. Its predictions have been confirmed to an accuracy of better than one part in ten billion. Every piece of modern technology, from semiconductors to lasers to MRI machines, operates on principles derived from quantum theory. It is not speculative. It is not approximate. It is, by every measurable standard, correct.

And nobody understands it.

This is not an exaggeration or a provocation. Richard Feynman, one of the greatest physicists of the twentieth century and a key architect of quantum electrodynamics, said plainly that nobody understands quantum mechanics. He was not being modest. He was being precise. The theory produces accurate predictions. The mathematics work. But the reality those mathematics describe is so far removed from human intuition that no one has been able to construct a coherent picture of what is actually happening.

Start with the basics. In classical physics, an object has a definite position and a definite velocity at any given moment. A ball is here, moving at this speed, in this direction. You can measure all of these properties simultaneously with as much precision as your instruments allow. The object has a state, and that state is real and definite regardless of whether you are looking at it.

In quantum mechanics, this is not the case. A particle does not have a definite position until it is measured. Before measurement, it exists in a superposition of all possible positions, described by a mathematical object called a wave function. The wave function does not tell you where the particle is. It tells you the probability of finding the particle in any given location if you choose to look. Before you look, the particle is not somewhere specific. It is, in a mathematical sense, everywhere it could possibly be, with varying degrees of likelihood.

This is not a limitation of measurement. It is not that the particle has a real position and we simply do not know what it is. The theory states, and experiments confirm, that the particle genuinely does not have a definite position before measurement. The uncertainty is not in our knowledge. It is in reality itself.

Werner Heisenberg formalized this in his uncertainty principle. You cannot simultaneously know the exact position and exact momentum of a particle. The more precisely you measure one, the less precisely you can know the other. This is not a technological problem that better instruments could solve. It is a fundamental property of nature. The universe does not contain the information you are asking for. It is not hidden. It does not exist.

This violates everything the human mind expects about reality. We live in a world of definite objects in definite places doing definite things. A chair is either in the room or it is not. A coin is either heads or tails. The idea that something can be genuinely undefined, not unknown but undefined, until the moment you observe it has no equivalent in everyday experience. The brain has no model for it. It can learn the mathematics. It can pass the exam. But it cannot form an intuitive picture, because no intuitive picture exists.

Then there is the measurement problem. When you observe a quantum system, the wave function collapses. The particle goes from being in a superposition of possible states to being in one definite state. This happens instantaneously. It is not a gradual process. It is a sudden transition from many possibilities to one actuality.

But what counts as a measurement? What is it about observation that causes the collapse? Is it the instrument? Is it the conscious observer? Is it any interaction with a larger system? Quantum mechanics does not say. The theory tells you that collapse happens when a measurement occurs. It does not define what a measurement is. This is not a minor omission. It is a hole in the foundation of the most successful theory in physics. The theory works perfectly in practice while containing a conceptual gap that has remained open for nearly a century.

Different interpretations of quantum mechanics attempt to fill this gap, and none of them are satisfying.

The Copenhagen interpretation, the oldest and most commonly taught, says that the wave function is not a description of reality. It is a tool for calculating probabilities. Do not ask what the particle is doing before you measure it. The question has no meaning. Just calculate and move on. This works operationally. It allows physicists to use the theory without getting tangled in philosophical knots. But it resolves the problem by declaring it off-limits, not by solving it.

The many-worlds interpretation says the wave function never collapses. Instead, every possible outcome of every quantum event occurs in a separate branch of reality. When you measure a particle and find it spinning clockwise, another version of you in another branch finds it spinning counterclockwise. Both outcomes are real. Both happen. The universe splits at every quantum interaction into an ever-expanding tree of parallel realities.

This is mathematically consistent. It eliminates the measurement problem entirely. But it does so at an extraordinary cost: an infinite number of unobservable parallel universes, none of which can ever be detected, confirmed, or communicated with. It is the ultimate unfalsifiable proposition. It explains everything by proposing something that can never be tested. Science producing a theory that science cannot evaluate.

Then there is quantum entanglement. Two particles can be prepared in a shared quantum state such that measuring one instantly determines the state of the other, regardless of the distance between them. This has been confirmed experimentally across distances of over a thousand kilometers. The correlation is immediate. Not fast. Immediate. There is no signal traveling between the particles. No force. No wave. The information, if you can call it that, appears to transfer instantaneously.

Einstein called this "spooky action at a distance" and believed it proved quantum mechanics was incomplete. He thought there must be hidden variables, predetermined properties that the particles carried with them, that would explain the correlation without requiring any instantaneous connection. In 1964, John Bell devised a theorem that could test this. Experiments have since confirmed, repeatedly and decisively, that local hidden variables cannot explain quantum entanglement. The correlations are real. The instantaneous connection, whatever it is, is real. And no one can explain the mechanism.

Science can describe entanglement with mathematical precision. It can predict the outcomes of entanglement experiments with perfect accuracy. But it cannot explain how two particles separated by a thousand kilometers coordinate their behavior instantaneously without any form of communication. The description is complete. The understanding is absent.

This is the deepest illustration of the limits of science. Not a case where science has failed. A case where science has succeeded, spectacularly, and the success itself reveals the boundary. Quantum mechanics works. It predicts. It delivers technology. It transforms the world. And at its core, it describes a reality that no human mind can comprehend. A reality where objects do not have definite properties until they are observed. Where observation itself alters what is being observed. Where particles separated by vast distances behave as if they are the same particle. Where the most fundamental layer of reality operates on rules that have no analogy in human experience.

Science found the rules. Science confirmed the rules. Science built a civilization on the rules. And science cannot tell you what the rules mean. It can tell you what happens. It cannot tell you why. It can tell you how to calculate the outcome. It cannot tell you what is happening between the calculations.

The limits of science are not failures. They are not signs that science should be replaced or abandoned. Nothing else comes close to its power, its precision, or its track record. But the limits are real. They are structural. And they are permanent, at least as far as current understanding can determine.

Science can describe the behavior of the universe with extraordinary accuracy. It cannot explain why the universe behaves that way. It can map the patterns. It cannot reach beneath them. It can answer "how" with breathtaking detail. It has almost nothing to say about "why."

And for the questions in this book, the questions about origins and boundaries and the nature of existence itself, "how" is not enough. "How" is a description. "Why" is where understanding would begin. And science, for all its power, stops at the threshold.

10

The Limits of Logic

If science has limits, then perhaps logic can go further. That is a reasonable assumption. Science depends on observation, and observation depends on instruments, and instruments have physical constraints. Logic has none of these dependencies. Logic operates purely in the realm of abstract reasoning. It does not need a telescope or a particle accelerator. It needs only a set of premises and a set of rules for drawing conclusions from those premises. It should, in theory, be able to reach places that science cannot.

And it can. Logic can address questions about the structure of arguments, the consistency of systems, and the validity of reasoning in ways that science never could. It is the backbone of mathematics, the foundation of computer science, and the framework within which every formal argument operates. Without logic, science itself would have no means of drawing conclusions from data. Logic is not just a complement to science. It is a prerequisite for it.

But logic, like science, has limits. And those limits are not minor. They are fundamental, proven, and inescapable. They do not arise from a lack of effort or cleverness. They arise from the nature of logic itself.

The most famous demonstration of this came in 1931, when a young Austrian mathematician named Kurt Godel published two theorems that permanently altered the landscape of formal reasoning. Godel's incompleteness theorems are among the most important results in the history of mathematics, and their implications extend far beyond mathematics into philosophy, computer science, and the foundations of knowledge itself.

The first incompleteness theorem states that in any consistent formal system that is capable of expressing basic arithmetic, there exist statements that are true but cannot be proven within that system. Not statements that are difficult to prove. Not statements that require more time or better techniques. Statements that are true and that the system will never, under any circumstances, be able to prove.

This was devastating. At the time, the prevailing ambition in mathematics was to build a complete, consistent formal system that could, in principle, prove every true mathematical statement. David Hilbert, one of the leading mathematicians of the era, had proposed this as the central goal of the discipline. Find a set of axioms. Derive all truths from them. Close the system. Make mathematics complete.

Godel proved this was impossible. Not impractical. Impossible. Any system powerful enough to do basic arithmetic will contain truths that it cannot reach. No matter how many axioms you add, no matter how you extend the system, there will always be statements that are true within the system and unprovable by the system. Completeness and consistency cannot coexist in any sufficiently complex formal framework.

The second incompleteness theorem goes further. It states that no consistent formal system capable of basic arithmetic can prove its own consistency. In other words, even if the system is consistent, it cannot demonstrate that fact using its own rules. To prove the system is reliable, you need a more powerful system. And that more powerful system, in turn, cannot prove its own consistency. The problem cascades upward infinitely.

This means that logic, at its most formal and rigorous, contains built-in blind spots. There are truths it can express but not verify. There are questions about its own reliability that it cannot answer from within. It is not a closed, self-sufficient system. It is an open one, with permanent gaps that cannot be filled by any extension or improvement of the system itself.

For most everyday purposes, this does not matter. Godel's theorems apply to formal systems of a specific type, and most practical reasoning does not bump up against their constraints. You can balance a budget, plan a trip, and construct a legal argument without ever encountering an undecidable proposition. The theorems are not relevant to the kind of thinking most people do on a daily basis.

But they are relevant to the kind of thinking this book is about.

The questions in this book are precisely the kind of questions that push against the limits of formal reasoning. What was before the beginning? What is outside everything? What is the nature of nothing? These are questions that operate at the boundary of what any system, logical, scientific, or philosophical, can process. And Godel showed that boundaries of this kind are not contingent. They are inherent. Every sufficiently complex system has them.

This has a direct implication for the project of answering ultimate questions through pure reason. If logic itself contains unprovable truths, then the hope that sheer reasoning can reach answers that observation cannot is misplaced. Logic can go further than science in some directions, but it eventually hits its own wall. And that wall is not made of ignorance or laziness. It is made of mathematics. It is proven. It is permanent.

Beyond Godel, there are other limits to logic that are less formal but equally important.

Logic operates on premises. Every logical argument begins with a set of assumptions, and the conclusion is only as reliable as those assumptions. If the premises are true and the reasoning is valid, the conclusion follows necessarily. But logic cannot verify its own premises. It can check whether a conclusion follows from a set of starting points. It cannot check whether the starting points are correct. That requires something outside of logic: observation, intuition, consensus, or another set of premises, which themselves require verification.

This means that every logical system is ultimately grounded in something non-logical. At the base of every chain of reasoning, there is an axiom, a starting assumption that is accepted without proof. In mathematics, these are explicit and acknowledged. In everyday reasoning, they are usually implicit and unexamined. But they are always there. Logic does not stand on its own. It stands on a foundation it cannot inspect.

This foundation problem is not academic. It affects every argument ever constructed about the fundamental nature of reality.

Consider the cosmological argument for the existence of God. Everything that exists has a cause. The universe exists. Therefore the universe has a cause. That cause is God. The logic is valid. If you accept the premises, the conclusion follows. But the first premise, that everything that exists has a cause, is an assumption. It is not proven. It is not self-evident. It is drawn from everyday experience, where things do appear to have causes, and then applied to the totality of existence, where everyday experience may not hold. Quantum mechanics, as discussed in the previous chapter, demonstrates that at the subatomic level, events can occur without deterministic causes. The premise may simply be wrong. And if the premise is wrong, the logic, no matter how valid its structure, produces nothing reliable.

The same applies in the other direction. Consider the argument that the universe needs no cause because time began with the Big Bang and causation requires time. The logic is again valid. If causation requires time, and time began with the universe, then asking for a cause of the universe is incoherent. But the premise that causation requires time is an assumption. It is drawn from our understanding of physics, which may be incomplete. There may be forms of causation that operate outside of time. We do not know. And because we do not know, the logical argument, however clean its structure, rests on ground that has not been secured.

This is the trap that logic falls into when applied to ultimate questions. The reasoning is rigorous. The structure is airtight. And the whole thing floats on premises that nobody can verify. It is a perfect building constructed on sand. Not because the builder was careless, but because sand is all there is at that depth.

There is another limit to logic that is less discussed but equally significant. Logic is binary. A statement is true or false. A conclusion follows or it does not. There is no middle ground, no ambiguity, no spectrum. This is what gives logic its power. It is precise. It is definitive. It eliminates vagueness.

But reality may not be binary. Quantum mechanics again provides the clearest example. A particle in superposition is not in state A or state B. It is in both, or neither, or some combination that has no classical equivalent. Applying binary logic to a system that is fundamentally non-binary produces results that are technically valid within the logical framework but may not correspond to what is actually happening. The logic works. The reality does not fit inside it.

Fuzzy logic and multi-valued logics have been developed to address this. These systems allow for degrees of truth rather than absolute true-or-false values. They are useful in engineering and computer science, particularly in systems that need to handle ambiguous or incomplete information. But they do not resolve the deeper issue. They are still formal systems with rules and axioms. They still have Godelian limits. They still rest on premises that cannot be self-verified. Expanding the framework does not eliminate the boundaries. It relocates them.

There is also the problem of self-reference. Logic struggles with statements that refer to themselves. The most famous example is the liar paradox: "This statement is false." If the statement is true, then it is false. If it is false, then it is true. The statement cannot be consistently assigned a truth value. It is not that we lack the skill to resolve it. It is that the structure of the statement makes resolution impossible within standard logic.

Godel's proof, in fact, relied on a form of self-reference. He constructed a mathematical statement that essentially said "this statement cannot be proven in this system." If the system could prove it, then it would be proving something false, making the system inconsistent. If the system could not prove it, then the statement was true but unprovable, making the system incomplete. Either way, the system fails to be both complete and consistent. Self-reference is not a trick or a curiosity. It is a structural feature that exposes a permanent crack in the foundation of formal reasoning.

This matters for the questions in this book because many of them are, at their core, self-referential. What is the nature of everything? The question includes itself. What are the limits of understanding? The question tests its own limits. What is nothing? The question attempts to define the absence of definition. Each of these turns inward at some point, and when they do, logic begins to stutter. Not because the question is too hard. Because the question folds back on itself in a way that logic was not designed to handle.

Mathematics, despite being the most rigorous expression of logical reasoning, has similar issues. There are mathematical problems that have been proven to be undecidable. Not unsolved. Undecidable. No algorithm can determine the answer. No proof can reach them. They exist within the system as permanent mysteries, visible but untouchable.

The most famous is the continuum hypothesis, which asks whether there is a set whose size is strictly between the size of the natural numbers and the size of the real numbers. In 1963, Paul Cohen proved that the hypothesis can neither be proven nor disproven from the standard axioms of set theory. It is independent of the system. The system can see the question but cannot answer it. Not now. Not ever. Not without changing the axioms, which would create a different system with its own undecidable questions.

This is the mathematical equivalent of the questions this book has been asking. A question that the system can formulate but not resolve. A question that sits inside the framework, visible and clear, permanently beyond reach.

Logic is the sharpest tool the human mind has ever built. It cuts through ambiguity, exposes contradictions, and forces clarity in a way that no other method of thinking can match. But it has edges it cannot sharpen, questions it cannot reach, and foundations it cannot inspect. It is powerful, precise, and incomplete.

The hope that pure reason could succeed where science and observation fall short is understandable. If the physical tools fail, perhaps the abstract ones will not. But Godel closed that door in 1931. Logic has blind spots. Mathematics has undecidable propositions. Formal systems have truths they cannot prove. The abstract tools are sharper than the physical ones, but they break on the same questions.

Every chapter in this book has ended at the same wall. Science reaches it through the limits of observation. Religion reaches it through the limits of explanation. And logic reaches it through the limits of proof. Three different paths. Three different methods. The same destination.

The wall does not move. It does not care which tool you bring. It is not waiting to be broken through. It may not be a wall at all. It may simply be the place where the ground ends and the mind, having run out of surface to walk on, insists that the path must continue because it cannot accept that it does not.

11

The Problem With "Why"

Why?

It is the first question a child learns to weaponize. Why is the sky blue? Because light scatters in the atmosphere. Why does light scatter? Because of the wavelength of blue light and the size of gas molecules. Why do molecules scatter shorter wavelengths more? Because of the physics of electromagnetic radiation. Why does electromagnetic radiation behave that way? Because of Maxwell's equations. Why are Maxwell's equations true?

Silence.

Every parent has experienced this. The chain of "why" that a child constructs, innocently, relentlessly, without any awareness of what they are doing, is one of the most efficient demonstrations of the limits of human knowledge ever devised. A four-year-old, armed with nothing but the word "why," can reduce a physicist to silence in under a minute. Not because the physicist is unintelligent. Because the chain of "why" always terminates at the same place: a description of how things are, with no explanation of why they are that way.

This is not a flaw in the child's reasoning. It is a flaw in the question. Or more precisely, it is a feature of the question that reveals something important about the structure of human knowledge. Every answer to a "why" question is, at best, a redirection to a deeper "why" question. And at the bottom of every chain, without exception, the redirections run out and all that remains is a brute fact. A thing that is the way it is, with no further explanation available.

Science is built on "how." How does gravity work? How do cells divide? How does light travel? These questions are productive. They lead to models, predictions, and technologies. They can be tested. They can be refined. They generate progress.

"Why" is different. "Why" asks for a reason, a purpose, a justification. It assumes that behind every phenomenon, there is not just a mechanism but a motivation. A cause in the deeper sense. Not just what made it happen, but what it is for. And this assumption, which feels so natural that most people never examine it, may be entirely unjustified.

The universe may not have a "why." It may only have a "how."

This is a difficult idea to sit with, because the human brain is a "why" machine. It does not just seek patterns. It seeks meaning. It does not just want to know what happened. It wants to know the reason it happened. This is deeply embedded in our cognitive architecture. Studies in developmental psychology show that children as young as three years old prefer explanations that include purpose over explanations that do not. When told that rocks are pointy, a three-year-old is more satisfied by "so that animals can scratch on them" than by "because bits of stuff piled up over time." The purposeful explanation is preferred not because it is more accurate, but because the brain is wired to find purpose more satisfying than mechanism.

This preference does not disappear in adulthood. It becomes more sophisticated, better disguised, wrapped in the language of philosophy or theology or science. But the underlying impulse is the same. The adult who asks "why does the universe exist?" is doing the same thing the child does when asking why the sky is blue. They are looking for a purpose behind the fact. A reason. A point.

And this is where the concept of god enters the picture. Not as a theological claim. Not as a matter of faith or doctrine. But as a cognitive pattern. A label the brain generates when the chain of "why" reaches its end and the mind refuses to accept that there is nothing more.

Consider how the concept has shifted throughout human history. For the earliest human societies, gods were everywhere. Thunder was a god. The river was a god. The harvest was a god. Disease was a god. Death was a god. Every phenomenon that could not be explained was assigned an agent, a conscious force that acted with intention. The question "why did the river flood?" received the answer "because the river god is angry." The question "why did the child die?" received the answer "because a spirit took them." These were not lazy answers. They were the product of a brain that could not accept purposeless events. Every effect needed a cause, and every cause needed an intention behind it.

As knowledge expanded, the gods retreated. Thunder was no longer a god once atmospheric electricity was understood. Disease was no longer a curse once bacteria and viruses were identified. The harvest was no longer a divine gift once agricultural science explained soil chemistry and weather patterns. Each advance in understanding removed a god from the equation. Not by disproving the god directly, but by providing a mechanical explanation that made the god unnecessary.

But the gods never disappeared entirely. They moved. They retreated to the edges of knowledge, to the places where "how" had not yet reached. And the largest, most persistent god of all settled at the very end of the chain: the origin of everything. Why is there something rather than nothing? God. Why does the universe exist? God. Why are the laws of physics what they are? God.

This is the god of the gaps. Not a specific deity with a name and a scripture. A cognitive function. The label the brain places on the boundary of its own understanding. Whatever the mind cannot explain, it calls god. And every time the boundary of explanation expands, god moves further back, occupying the next unanswered question, the next gap in the chain.

This is not limited to religious thinking. It happens in every framework.

In a quantum mechanical framework, the observer plays a godlike role. The act of observation collapses the wave function. It determines which possibility becomes real. Without observation, the system exists in superposition, undefined, uncommitted. The observer is the agent that forces reality into a definite state. That is not a religious claim. It is a description of the theory. But the structural parallel is striking. An undefined reality that requires a conscious agent to become real. Replace "observer" with "creator" and the sentence could appear in a religious text.

In simulation theory, the programmer is god. The programmer designs the rules, sets the parameters, initiates the system, and exists outside of it. The simulated beings cannot access the programmer. They cannot verify the programmer's existence from within the simulation. They can only infer, based on the structure of their reality, that something or someone must have built it. Replace "programmer" with "God" and you have the argument from design, restated in technological language.

In animist and pagan traditions, nature itself is god. The forest, the ocean, the cycle of seasons, the turning of the earth. Divinity is not a separate being that created nature. Divinity is nature. The sacred is embedded in the physical world, not above it. This is a different concept of god, but it serves the same function. It answers the question "why" by pointing to something and declaring it ultimate. Why does the world exist? Because nature is divine. Why is nature divine? The question is not permitted. The chain stops.

In Wiccan and similar traditions, the divine is often understood as a duality, a god and a goddess, representing complementary forces. Creation and destruction. Growth and decay. Light and darkness. The world exists because these forces interact. This is an elegant framework. It maps onto observable reality with surprising precision. But it does not escape the same structural limitation. Why do these forces exist? Why these two and not others? The framework explains the pattern of the world by attributing it to divine forces, but it cannot explain the forces themselves.

For a small animal, a pet mouse or a dog, the human being who provides food, controls the environment, and determines the boundaries of the animal's existence functions as a god. The animal cannot comprehend the human's nature. It cannot understand the human's motivations. It cannot access the larger world the human inhabits. It knows only that there is a being of immense power whose actions shape its entire reality. It does not understand the being. It simply responds to it.

This is not a trivial observation. It is a structural model. The relationship between a pet and its owner may be, in terms of information asymmetry and power differential, the closest available analogy for the relationship between a human being and whatever, if anything, sits beyond the boundary of human understanding. We are the mouse. We sense that something shapes our reality. We cannot comprehend it. We respond to it with stories, theories, rituals, and equations. And we call it by whatever name our framework provides.

The word "god" does not describe a fixed entity. It describes a function. It is the label the mind attaches to the edge of its own comprehension. And that edge moves. For early humans, the edge was close. Almost everything was beyond it. As knowledge expanded, the edge retreated. Fewer things needed the label. But the edge never disappeared. It simply moved to bigger, harder, more abstract questions. And at every stage, whatever sat just beyond the edge was called god.

This is why the question "why" is so problematic. It assumes that there is a reason to be found. It assumes that behind every mechanism, there is a purpose. And when the chain of mechanisms runs out, the mind places a god at the end, whatever form that god takes, because it cannot accept that the chain simply stops.

But what if the chain does simply stop?

What if "why" is a question that only functions within a certain range, the way Newtonian physics only functions at certain scales? What if "why" works perfectly well for everyday events, why did the car crash, why did the business fail, why did the relationship end, but breaks down completely when applied to the fundamental structure of reality?

There is no empirical evidence that the universe has a purpose. There is no observation that confirms the existence of a reason behind the laws of physics. There is no measurement that detects intention in the structure of matter. The assumption that there is a "why" behind everything is a projection of human cognition onto a universe that has never confirmed it. We ask "why" because our brains are built to ask "why." The universe does not appear to be built to answer it.

This does not mean the universe definitely has no purpose. Absence of evidence is not evidence of absence, as the saying goes. But the distinction matters less than it appears. Whether the universe has a purpose we cannot detect, or has no purpose at all, the practical result is the same. We do not have access to it. We cannot confirm it. We cannot build on it. We are left in exactly the same position either way, holding a question that produces no verifiable answer regardless of which possibility is true.

The problem with "why" is not that it is a bad question. It is an extraordinary question. It has driven the entire enterprise of human thought for thousands of years. But it carries a hidden assumption that may not be valid, the assumption that reasons exist at every level of reality. And when that assumption is applied to the deepest questions, the ones about origins and boundaries and the nature of existence itself, it produces a chain that ends not in an answer but in a label. God. Nature. The programmer. The observer. The quantum field. Each one a different name for the same thing: the place where "why" stops working and the mind, unable to accept the silence, fills it with a word.

The child who asks "why" five times in a row is not being difficult. The child is performing, in miniature, the entire history of human inquiry. Each answer leads to a deeper question. Each deeper question leads to a more fundamental answer. And at the end, the parent says "because that is just the way it is." The child is unsatisfied. The parent is exhausted. And the question remains, hanging in the air between them, unanswered and possibly unanswerable.

Every philosopher, every theologian, every scientist who has ever lived has been that child. And every framework they have built has been a more sophisticated version of the parent's final answer. This is just the way it is. We call it God. We call it the laws of physics. We call it the fundamental nature of reality. But the structure is the same. A declaration that the chain ends here, dressed in language that makes the ending feel intentional rather than forced.

The problem with "why" is that it never accepts that answer. It always wants one more step. And there is always one more step available, until there is not. And when there is not, the mind does not rest. It builds a god to stand in the gap. And it has been doing this for as long as minds have existed.

The question is not whether gods are real. The question is whether "why" is the right tool for reaching them. And the answer, based on everything we have examined, is that "why" is a tool that builds its own destination. It does not find the answer. It constructs it. And what it constructs tells you more about the mind that built it than about the reality it claims to describe.

12

Questions That Cannot Exist

Every chapter so far has dealt with questions that can be asked but not answered. This chapter is about something stranger. Questions that cannot be asked. Not because they are forbidden or taboo, but because the structure of language and thought does not permit them to form.

This sounds paradoxical. If a question cannot exist, how can you talk about it? If it cannot be formed, how can you point to the space where it would be? The answer is that you can approach the edges of these questions the same way you can approach the edge of a cliff. You can stand close enough to see that the ground ends. You cannot step into the air beyond it. But you can look down and recognize that something is there, or rather, that something is not there, in a way that matters.

Start with a simple observation. Language is finite. The number of words in any language is finite. The number of grammatical structures is finite. The number of possible sentences, while enormous, is bounded by the rules of syntax and the limits of human cognition. This means that the number of questions that can be formed in any human language is, in principle, limited. It is an astronomically large number, but it is a number. It is not infinite.

Reality, on the other hand, may be infinite. Or it may be structured in ways that do not map onto the categories embedded in human language. Either way, there is no guarantee that the set of questions humans can ask covers the set of things that are true about reality. There may be aspects of existence that no question in any language can reach. Not because the question is too complex to articulate, but because the conceptual framework required to form it does not exist in any human language or any human mind.

This is not a speculative claim. There are concrete examples.

Consider a being that experiences a dimension of reality you do not have access to. Not a spatial dimension in the way we normally discuss them, but an entirely different category of experience that has no equivalent in human perception. You cannot form a question about this dimension because you have no concepts to build the question from. You do not lack the vocabulary. You lack the perceptual and cognitive foundation. A person born without the ability to see cannot form the question "what does blue look like to a person with synesthesia?" because the question requires access to a perceptual category that does not exist in their experience. They can learn the words. They can memorize the sentence. But they cannot form the question in the way a sighted person can, because the underlying concept is absent from their cognitive architecture.

Now scale this up. Human beings experience three spatial dimensions and one temporal dimension. Our entire conceptual framework, every word, every idea, every question we have ever formed, is built on this foundation. If reality contains aspects that do not map onto these dimensions, and there are theoretical reasons to believe it might, then there are questions about those aspects that we cannot form. Not because we are not smart enough. Because the raw materials for constructing those questions do not exist in our minds.

This is different from the limits discussed in previous chapters. The limits of science are about what can be observed and tested. The limits of logic are about what can be proven and derived. The limits discussed here are about what can be conceived. They are pre-scientific and pre-logical. They operate at the level of thought itself, before reasoning even begins.

Ludwig Wittgenstein addressed this directly. His famous proposition, that the limits of my language are the limits of my world, was not a poetic statement. It was a claim about the structure of thought. If you cannot express something in language, you cannot think it. If you cannot think it, you cannot ask about it. The boundaries of language are the boundaries of inquiry. Whatever lies outside those boundaries is not just unknown. It is unthinkable.

This has been debated extensively. Some philosophers argue that thought can exceed language, that there are experiences and intuitions that resist verbal expression but are still real cognitive events. The feeling of deja vu. The sense of understanding something you cannot put into words. The tip-of-the-tongue phenomenon, where you know something but cannot articulate it. These suggest that cognition is broader than language, that the mind can reach places that words cannot follow.

But even if this is true, it does not solve the problem. It merely pushes the boundary from language to cognition. If there are aspects of reality that exceed not just language but cognition itself, then the boundary still exists. It is just further out. The space beyond it is still inaccessible. And the questions that would address that space still cannot be formed, whether in words or in thought.

Consider the ant. An ant navigates its world with a set of senses and cognitive tools appropriate to its scale. It can detect chemical trails, temperature gradients, and physical obstacles. It can communicate with other ants through pheromones. It operates effectively within its environment. But there are questions about the ant's world that the ant cannot ask. It cannot ask about the geopolitical situation of the country it lives in. It cannot ask about the chemical composition of the pheromones it follows. Not because these questions are too difficult for it, but because the conceptual architecture required to form them does not exist in an ant's brain. The ant does not know that it does not know. The questions are not hard for the ant. They do not exist for the ant.

Now apply this to human beings. We are more cognitively complex than ants by an enormous margin. But we are not infinitely complex. Our brains have a finite number of neurons, a finite number of connections, a finite processing capacity. There is no reason to assume that our cognitive architecture is sufficient to form every possible question about reality. There may be aspects of existence that are to us what geopolitics is to an ant. Not difficult. Not hidden. Simply outside the range of what our minds can construct.

This is a deeply disorienting idea, because it means that the set of questions we can ask is not the same as the set of questions that exist. There are questions we are missing. And we cannot know what they are. We cannot even know how many there are. We cannot feel the edges of this gap because the gap is defined by the absence of the cognitive tools that would allow us to detect it.

There is a temptation, when confronted with this idea, to dismiss it as abstract philosophy with no practical relevance. The questions we can ask seem sufficient. Science progresses. Technology advances. Understanding deepens. If there are questions we cannot form, they do not seem to interfere with the questions we can.

But this is the same reasoning the ant would use if the ant could reason. Everything in the ant's world is accounted for. The chemical trails work. The colony functions. The food arrives. From inside the ant's cognitive framework, nothing is missing. The absence of higher-order questions does not register as an absence. It registers as completeness. The ant's world feels whole because the ant has no concept of what is missing from it.

The human equivalent is more subtle but structurally identical. Our world feels whole. Our questions seem comprehensive. The gaps we are aware of, the questions we know we cannot answer, feel like they represent the full extent of our ignorance. But that feeling is an artifact of the cognitive system itself. We can only perceive the gaps that fall within the range of our perception. Gaps that fall outside that range do not appear as gaps. They do not appear at all.

This is related to a concept in epistemology known as unknown unknowns. Donald Rumsfeld, speaking in a political context, distinguished between known knowns, known unknowns, and unknown unknowns. The known knowns are things we know we know. The known unknowns are things we know we do not know. The unknown unknowns are things we do not know that we do not know. The third category is the most dangerous, not because it contains the most difficult questions, but because it contains questions we have not yet recognized as questions. And the category discussed in this chapter goes further still. It is not about questions we have not yet asked. It is about questions we cannot ask. Questions that the structure of human cognition prevents from forming.

Mathematics provides a useful parallel. In any formal system, as Godel demonstrated, there are true statements that the system cannot prove. The system cannot identify these statements from within. It cannot point to them and say "here is something I cannot prove." It simply cannot reach them. They are invisible from the inside. A mathematician working within the system would have no way of knowing what is missing, because the tools required to detect the gap are the same tools that are insufficient to bridge it.

Human cognition may be in the same position. There may be truths about reality, important ones, fundamental ones, that the human brain cannot formulate as questions. Not because the brain is damaged or underdeveloped. Because the brain is a specific organ with a specific architecture designed for a specific range of tasks. It is extraordinarily powerful within that range. Outside it, there is no guarantee it functions at all.

Some thinkers have attempted to push against this boundary by building tools that extend cognition. Mathematics itself is one such tool. It allows the mind to manipulate abstractions that it could not handle through intuition alone. The concept of eleven-dimensional space in string theory, for example, is not something any human can visualize. But it can be expressed and manipulated mathematically. The math serves as a prosthetic for a cognitive limitation.

Computers extend this further. They can process calculations that no human brain could perform, identify patterns in data sets too large for human analysis, and simulate systems too complex for human intuition. Artificial intelligence, in its current and projected forms, promises to extend cognitive reach even further, potentially identifying questions that human minds could not generate on their own.

But there is a limit to this extension as well. Every tool humans build is built from human concepts. Every question a computer asks was either programmed by a human or derived from a framework that a human designed. The tool inherits the conceptual boundaries of its creators. A more powerful calculator can solve harder equations. It cannot ask questions that fall outside the mathematical framework it operates within. A larger telescope can see more distant objects. It cannot observe phenomena that do not interact with light. Extension is not the same as transcendence. Making the existing tools more powerful does not create new categories of thought. It makes the existing categories more productive.

This is why artificial intelligence, despite its extraordinary capabilities, may not resolve the problem discussed in this chapter. An AI that operates on human-designed principles, trained on human-generated data, using human-created mathematics, is an extension of the human cognitive framework. It is not a departure from it. It can go further within the framework. It is not clear that it can go outside it. And the questions that cannot exist are, by definition, outside it.

There is something vertiginous about this realization. It means that human knowledge, no matter how vast it becomes, may be permanently incomplete in ways that we cannot detect. Not incomplete in the sense that there are questions we have not yet answered. Incomplete in the sense that there are questions we have not yet conceived and cannot conceive. The map of human inquiry, no matter how detailed it becomes, may be missing entire continents that no instrument can reveal and no mind can imagine.

This is the deepest limit explored in this book. The previous chapters dealt with questions that push against the boundaries of science, logic, and language. This chapter deals with the boundaries themselves. Not the edges you can see when you look out from the shore. The edges you cannot see because you do not know the shore has a shape.

Every question in this book has been a question that the human mind can form but not answer. The possibility raised here is that these are not the hardest questions. They are the hardest questions we can see. Beyond them, in a space we cannot access, there may be questions that would make every unanswered question in this book look simple. Questions that would reframe everything we think we know, if only we could think them.

We cannot. And we cannot know what we are missing. The only thing we can know is that the claim of completeness, the feeling that our questions cover all of reality, is almost certainly an illusion. The brain feels like it can ask about everything. It probably cannot. And the gap between what it can ask and what exists is not a gap we can measure, approach, or ever fully acknowledge.

This is the quietest form of the wall this book keeps encountering. Not a wall you hit. A wall you cannot see. Not a question without an answer. An answer without a question. Something true about reality that has no corresponding thought in any human mind. Not yet. And possibly not ever.

The previous chapters ended with unanswered questions. This chapter ends with the possibility that the most important questions have not been asked. That they sit in a space the mind cannot enter, in a language the brain does not speak, describing aspects of reality that no human framework can represent.

That is not frustrating in the way the other chapters are frustrating. It is something quieter and more permanent. A silence you cannot hear because you do not have the ears for it.

13

The Observer Problem

Who is watching?

This is not a paranoid question. It is a physics question. And it is one of the strangest, most unsettling problems in the history of science. Because in quantum mechanics, the act of watching changes what is being watched. Not metaphorically. Not in the way that a camera might make someone nervous at a party. Literally. Physically. Measurably. The presence of an observer alters the outcome of an experiment at the most fundamental level of reality.

The double-slit experiment is the clearest demonstration of this, and it has been called the most beautiful experiment in physics. The setup is simple. Fire particles, photons or electrons, one at a time, at a barrier with two narrow slits. On the other side of the barrier, place a detector screen that records where each particle lands. Run the experiment long enough and a pattern emerges.

If the particles were behaving like tiny balls, you would expect two clusters on the screen, one behind each slit. That is not what happens.

Instead, the screen shows an interference pattern, alternating bands of high and low density, exactly the pattern you would expect if waves were passing through the slits and interfering with each other. Peaks where the waves reinforce. Troughs where they cancel. The particles, fired one at a time with no possibility of interacting with each other, produce a wave pattern.

This means that each individual particle, somehow, passes through both slits simultaneously and interferes with itself. It behaves as a wave, spread across multiple paths, until it hits the screen and registers as a single point. A wave in transit. A particle on arrival.

This is already strange enough. But it gets stranger.

If you place a detector at the slits to determine which slit each particle passes through, the interference pattern disappears. The particles behave like ordinary objects, passing through one slit or the other, producing two clusters on the screen. No interference. No wave behavior. The simple act of observing which path the particle takes changes the result of the experiment.

Remove the detector. The interference pattern returns. Add the detector. It disappears. The particles seem to know whether they are being watched and adjust their behavior accordingly.

This has been confirmed thousands of times across decades of experimentation. It is not disputed. It is not a fringe result. It is one of the most replicated findings in the history of physics. And no one can fully explain it.

The standard interpretation is that observation causes the wave function to collapse. Before observation, the particle exists in a superposition of all possible states. It does not have a definite path. It does not go through one slit or the other. It goes through both. It goes through neither. It exists as a probability distribution, a mathematical description of all the things it might do. When you observe it, the superposition collapses into a single definite state. The particle becomes a particle. The wave becomes a point.

But what is observation? What qualifies? This is where the problem deepens.

In the original Copenhagen interpretation, observation was loosely associated with measurement by a macroscopic device. The act of recording information about the system, using any physical apparatus, was sufficient to cause collapse. The observer did not need to be conscious. A camera would do. A Geiger counter would do. Any interaction between the quantum system and a larger, classical system counted as observation.

But this raises an immediate question. If a Geiger counter can cause wave function collapse, what makes the Geiger counter special? It is made of atoms. Those atoms are governed by quantum mechanics. The Geiger counter itself should be in superposition until something observes it. And whatever observes the Geiger counter should also be in superposition until something observes that. The problem does not stop at the first observer. It cascades outward in an infinite regression of observers observing observers.

This is known as the Von Neumann chain, and it terminates, in Von Neumann's original formulation, at the conscious mind. The chain of quantum systems interacting with quantum systems only collapses when a conscious being becomes aware of the result. Consciousness, in this interpretation, is not just a participant in reality. It is the mechanism by which reality becomes definite.

This is an extraordinary claim. It places consciousness at the center of physics, not as a side effect of material processes but as a fundamental ingredient in the determination of physical reality. Without an observer, reality remains undefined. Without consciousness, the wave function never collapses. The universe, in the absence of a mind to observe it, exists only as a cloud of probabilities, never committing to a definite state.

Not all physicists accept this interpretation. Many reject it explicitly. The many-worlds interpretation avoids the problem entirely by stating that the wave function never collapses. Every possible outcome occurs in a separate branch of reality. There is no need for an observer because nothing is selected. Everything happens. The appearance of collapse is an illusion created by the fact that each branch can only see itself.

The decoherence program offers another approach. It explains the appearance of collapse as the result of quantum systems interacting with their environment. When a quantum particle interacts with a large number of other particles, its quantum properties become entangled with the environment and effectively disappear from the perspective of any local observer. The particle appears to be in a definite state, not because anything special happened, but because the quantum information has spread out into the environment and become practically inaccessible.

Decoherence is well supported experimentally. It explains why quantum effects are not visible at the macroscopic scale. It explains why your coffee cup does not exist in a superposition of positions. But it does not fully solve the measurement problem. It explains why superpositions become unobservable. It does not explain why one particular outcome occurs rather than another. The selection of a specific result from among all possible results remains unexplained. Decoherence tells you why you see a definite world. It does not tell you why you see this definite world rather than a different one.

The observer problem is not just a technical puzzle in physics. It is a question about the relationship between consciousness and reality.

And that question, depending on how you approach it, leads to some of the most disorienting conclusions in all of human thought.

If consciousness is required for wave function collapse, then consciousness is not a product of the physical world. It is a prerequisite for it. The physical world does not produce consciousness. Consciousness produces the physical world. Or, at the very least, consciousness and the physical world are co-dependent, each requiring the other to exist. Neither comes first. Neither is more fundamental. They arise together, or not at all.

This inverts the standard materialist view of reality, which holds that matter is fundamental and consciousness is a byproduct. In the materialist framework, atoms arranged in sufficiently complex patterns produce brains, and brains produce consciousness. Consciousness is late to the story. It is an effect, not a cause. The universe existed for billions of years before any conscious being appeared to observe it. Stars formed, galaxies collided, planets coalesced, all without anyone watching.

But if observation is necessary for reality to become definite, then what was the universe before anyone observed it? Was it real? Was it a vast, formless probability cloud, waiting for the first conscious being to look at it and force it into a definite state? Did the Big Bang happen if no one was there to see it?

These sound like absurd questions. They are not. They are direct consequences of taking the observer's role in quantum mechanics seriously. John Wheeler, one of the most respected physicists of the twentieth century, proposed a thought experiment called the delayed-choice experiment, which was later confirmed in the laboratory. In this experiment, the decision of whether to observe a particle's path can be made after the particle has already passed through the slits.

And the result still changes. The particle's past behavior is altered by a future observation.

This does not mean information travels backward in time. The interpretation is more subtle and more disturbing. It suggests that the past is not fixed until it is observed. That what happened depends, in some fundamental way, on what is measured later. The history of the particle is not determined when the particle moves. It is determined when the observation is made. Wheeler extended this idea to the cosmic scale. He suggested that observations made today could, in a quantum mechanical sense, participate in shaping the conditions of the early universe. That the observer is not a passive witness to a pre-existing history but an active participant in bringing that history into being.

This is not mysticism. This is theoretical physics, published in mainstream journals, tested in laboratories, and debated by serious scientists. It is uncomfortable not because it is unscientific but because it challenges the most basic assumption we hold about reality: that it exists independently of us. That the universe was there before we arrived and will be there after we are gone. That observation is passive. That we are witnesses, not participants.

The observer problem suggests that this assumption may be wrong. Not that the universe is imaginary. Not that reality is a figment of consciousness. But that the relationship between the two is far more entangled than common sense suggests. The observer does not sit outside reality and look in. The observer is embedded in reality, and reality is, in some way that we do not fully understand, shaped by the act of observation.

This connects directly to the question of god, as discussed in the previous chapter. If the observer plays a fundamental role in determining reality, then the observer functions, at the quantum level, as a creator. Not in the grand, theological sense. But in the precise, physical sense that the act of observation selects which possibility becomes actual. The observer decides, without intention or awareness, which version of events becomes real. That is a godlike function performed by every conscious being every moment of every day. Not through will. Not through power. Through the mere act of perceiving.

This does not prove that consciousness is god. It does not prove anything about god at all. But it demonstrates that the boundary between observer and creator is not as clear as we assume. In classical physics, the observer is irrelevant. The universe does what it does whether anyone is watching or not. In quantum mechanics, the observer is woven into the fabric of events. Remove the observer and the events are undefined. The observer is not watching the show. The observer is part of the stage.

Some physicists dismiss the significance of this by pointing out that "observer" in quantum mechanics does not mean a conscious being. It means any interaction that extracts information from a quantum system. A photon bouncing off an electron counts as observation. A particle detector counts as observation. Consciousness is not required. The wave function does not care whether a human is watching or a machine is recording.

This is technically correct within certain interpretations. But it does not eliminate the problem. It relocates it. If observation is just physical interaction, then every particle in the universe is constantly observing every other particle. The universe is a web of mutual observation, with every component constantly collapsing the wave functions of every other component. In this picture, the universe observes itself into existence. It is both the observer and the observed. It does not need an external consciousness to become real. It becomes real through its own internal interactions.

This is coherent. It is internally consistent. But it raises its own questions. If the universe observes itself, what was it before the first interaction? If observation is physical interaction, and physical interaction requires particles, and particles require observation to become definite, then you have a circular dependency. Observation requires particles. Particles require observation. Neither can come first. Neither can exist without the other. The system bootstraps itself into existence from a state that, by its own rules, should not have been possible.

This is the observer problem at its deepest. Not a question about laboratory equipment or experimental design. A question about the foundations of reality itself. Does the universe exist independently of observation? If so, quantum mechanics needs a new interpretation, because the current ones all assign the observer a role that cannot be removed without changing the theory. If not, then reality is not what we think it is. It is not a fixed stage on which events play out. It is a participatory process in which the observer and the observed are inseparable.

Neither option is comfortable. Neither option has been confirmed. Neither option leads to a clean, satisfying conclusion.

The observer problem is, in many ways, the purest expression of everything this book has been exploring. A question that science can pose but not resolve. A problem that logic can frame but not close. A mystery that sits at the intersection of physics, philosophy, and consciousness, touching all three and belonging fully to none.

Who is watching? The question assumes a separation between the watcher and the watched. Quantum mechanics suggests that separation may not exist. The observer is not outside the system. The observer is the system. And the system cannot fully observe itself, for the same reason an eye cannot see itself without a mirror.

We are the universe asking questions about itself. And the deepest question it can ask is whether it exists when it is not asking. That question has no answer. Not because the answer is hidden. But because the question, the act of asking, is itself part of the process that creates the thing being asked about.

14

The Final Question

What if we are a virus?

Not in some poetic sense. Not as a metaphor for environmental destruction or human greed. In a literal, structural sense. What if human beings are a viral infection inside a cell called Earth, inside a body so vast that we cannot detect it, in a world we will never see?

Look at what we do. We arrived on this planet. We spread across its surface. We consume its resources. We multiply. We alter its chemistry. We raise its temperature. We strip its surface. We burrow into it and extract material from deep inside it. We have changed the composition of its atmosphere, its water, and its soil. We have driven thousands of other species to extinction. We have spread to nearly every corner of the surface and we show no signs of stopping.

That is what a virus does to a cell.

A virus enters a cell. It hijacks the cell's machinery. It uses the cell's resources to replicate itself. It alters the cell's internal environment. It damages the cell's structures. It spreads until the cell either fights it off, adapts to it, or dies. The virus does not intend to destroy the cell. It does not have intentions. It simply does what it does, and the result is damage.

Now look at Earth. It is a sphere with a membrane of atmosphere. It has internal processes: tectonic movement, ocean circulation, atmospheric chemistry, carbon cycles. It maintains a temperature range. It regulates its own systems through feedback loops. It has been doing this for over four billion years. It functions. Not like a machine. Like a cell. A self-regulating, self-maintaining unit within a larger system.

And we are inside it. Billions of us. Consuming. Replicating. Altering. Damaging. But not enough to kill it. Not yet. The cell is stressed. It is running a fever. Its chemistry is shifting. But it is still functional. Still turning. Still regulating, as best it can, around the infection.

This is not a theory. It is not a scientific model. It cannot be tested or falsified. But that is the point. It cannot be disproven either. Nothing in physics, biology, or cosmology rules it out. Nothing says our universe cannot be a structure inside a larger structure inside a larger structure still. Nothing says Earth cannot be a functional unit inside an organism we cannot perceive. Nothing says we cannot be, in every meaningful sense, a pathogen.

Consider scale. A virus is roughly one hundred nanometers in diameter. A human cell is roughly ten micrometers. The ratio is about one to a hundred. A virus inside a cell has no way of perceiving the body the cell belongs to. It has no sensory apparatus for it. It has no cognitive framework for it. The concept of a body does not exist at the viral scale. The virus interacts with the cell and only the cell. The rest of reality, from the virus's perspective, does not exist.

Now scale upward. A human being is roughly 1.7 meters tall. The Earth has a diameter of roughly 12,700 kilometers. The ratio is enormous but finite. We interact with the Earth and, through our instruments, with a small portion of the space around it. We have sent probes to other planets. We have observed distant galaxies. But our interaction with the universe beyond our immediate environment is limited to observation. We do not touch it. We do not alter it. We look at it, and we interpret what we see through frameworks built at our scale.

What if those frameworks are as limited as a virus's interaction with its cell? What if everything we see when we look out at the universe is the equivalent of a virus detecting chemical gradients inside a cell membrane? Real data. Genuine information. Completely insufficient to describe the actual nature of the larger system.

The virus detects molecules. It responds to chemical signals. It processes information at its scale and acts accordingly. If a virus could think, it would build a model of its world based on those chemical signals. That model would be internally consistent. It would make accurate predictions within the cell. It would be, by the virus's standards, good science. And it would tell the virus absolutely nothing about the person whose body it inhabits. Nothing about the city that person lives in. Nothing about the planet. Nothing about the universe.

Not because the virus is stupid. Because scale imposes absolute limits on what can be known from within.

We have built extraordinary instruments. Telescopes that can see thirteen billion light years in every direction. Particle accelerators that can probe the structure of matter at scales smaller than a proton. Computers that can simulate the behavior of billions of particles interacting according to the laws of physics. Our science is, by any measure, remarkable.

But it is all built at our scale. By organisms of our size. Using materials available in our cell. Processing information through brains that evolved inside this specific environment. Every theory we have ever produced, every equation, every model, every framework, is the product of an infection trying to understand the cell it lives in using tools made from the cell's own material.

What if that is all it can ever be?

What if the questions in this book, every single one of them, are the equivalent of a virus trying to understand the body it inhabits by analyzing the chemistry of the cell wall? The data is real. The analysis is rigorous. The conclusions are valid within the frame. And the frame is so small, relative to what actually exists, that the conclusions are meaningless outside it.

If we are a virus inside a cell, then the body we inhabit is, by every functional definition, a god. Not a god that created us intentionally. Not a god that watches over us. Not a god that cares. A god in the structural sense. An entity so vast, so far beyond our scale, that it shapes every aspect of our existence without knowing we are here. The way you are a god to the bacteria on your skin right now. You control their environment. Your temperature, your sweat, your movement, the soap you use, the clothes you wear, all of it determines whether they live or die. You are their universe. And you do not know their names. You do not know they exist as individuals. You do not think about them. They are beneath your threshold of attention.

That is what god might be. Not a conscious creator. Not a designer. Not a father figure or a judge. Just a body going about its day, unaware that inside one of its cells, billions of tiny organisms have built cities, written books, launched wars, and spent thousands of years arguing about whether the body exists.

The body does not care about the argument. The body is eating lunch.

Consider time. A human lifespan is roughly eighty years. A bacterium can reproduce in twenty minutes. In a single human day, hundreds of generations of bacteria are born, live, and die. Entire lineages rise and fall between breakfast and dinner. What feels like an epoch to the bacterium is an unremarkable Tuesday to the human.

Scale this to our situation. If we are organisms inside a vastly larger system, then our sense of time is local. Our billions of years might be days to the body. The entire history of the universe, from the Big Bang to this moment, might correspond to a single phase in the body's life. Childhood, perhaps. Our 13.8 billion years of cosmic history might be the body growing up. And we, inside one of its cells, have measured that growth, mapped it, modeled it, and called it the expansion of the universe.

The universe is expanding. We know this. We have measured it. The rate of expansion has been calculated with extraordinary precision. Galaxies are moving apart. Space itself is stretching. We have built an entire cosmological framework around this observation.

But a body grows. A child's body expands. Its cells move apart. New cells form. The distances between structures increase. From inside a cell, during a period of growth, the environment would appear to be expanding. The cell membrane stretches. The space between organelles increases. Neighboring cells drift further away. If you were a virus inside that cell, with instruments capable of measuring the distances between structures, you would observe expansion. You would measure it. You would develop equations to describe it. You would find that the rate of expansion appears to be accelerating, because growth is not linear. A child grows faster at some stages than others.

You would call it cosmology. You would call it the greatest discovery of your civilization. And you would be describing a child getting taller.

The Big Bang, in this framework, is conception. The moment the body began. A single cell, dividing, multiplying, differentiating, building structures of increasing complexity. From inside, it looks like an explosion of matter and energy from a single point. From outside, it is just a life starting. An ordinary event in an ordinary world that does not know or care that the organisms inside one of its cells have given the event a name and spent centuries arguing about what preceded it.

What preceded it was the body's parents. What preceded the body's parents was their parents. The infinite regression that every chapter of this book has encountered, the endless chain of "but what came before that," might have a simple answer. It is just generations. Lives beginning and ending. Bodies growing and dying. An ordinary biological chain, unremarkable in the world where it occurs, incomprehensible from inside one of the cells.

And what about our space probes? We have sent machines to the Moon, to Mars, to Jupiter, to the edge of the solar system. We are proud of this. It represents the furthest reach of our species. We have touched other worlds.

A virus spreads from one cell to another. It crosses the membrane. It enters a new environment. It begins replicating in a new host cell. From the virus's perspective, if it had one, this would be an extraordinary achievement. A journey to a new world. An expansion of its civilization. From the body's perspective, it is an infection spreading. One more cell compromised. Not an achievement. A symptom.

Our probes on Mars might be exactly that. Not exploration. Infection. The virus reaching for the next cell. And if we eventually colonize Mars, which we discuss as the great next step for humanity, we will have done what every successful virus does. We will have spread beyond our original host cell and begun damaging a new one.

The body, if this framework holds, will eventually die. Everything does. And when it dies, the cells die with it. The chemistry stops. The temperature drops. The processes that maintained the environment inside each cell cease to function. The infection ends. Not because it was defeated. Not because the immune system won. Because the host stopped existing.

From inside the cell, this would look like the end of the universe. The fundamental processes that sustain reality would slow down and stop. Energy would dissipate. Structures would break apart. Activity would cease. We have a model for this. We call it heat death. The universe expanding into a cold, inert state where nothing happens and nothing can happen. The final equilibrium.

Or maybe the body does not die of old age. Maybe it gets hit by something. A sudden, catastrophic event. From inside the cell, this would look like the universe ending violently. A collapse. A tear in the fabric of space. An event without warning and without explanation. We have models for that too. The Big Crunch. The Big Rip. We have given names to scenarios that, from outside the cell, might just be a body having an accident on a Wednesday afternoon.

None of this can be proven. None of it can be tested. None of it can be confirmed or denied by any instrument we have or any instrument we could build. And that is precisely the point. Not that this framework is true. But that it is possible. Nothing in our knowledge rules it out. Nothing in physics, biology, chemistry, or any other discipline can say with certainty that we are not an infection inside a cell inside a body in a world we cannot see.

And if we cannot even rule out this possibility, then we cannot verify the foundation on which every question in this book has been asked. We have spent thirteen chapters examining the biggest questions human beings can conceive. The origin of the universe. The nature of matter. The meaning of nothing. The limits of science, logic, and thought. We have explored quantum mechanics, simulation theory, the observer problem, and the boundaries of human cognition. Every chapter hit the same wall. Every question went unanswered.

But maybe the questions were never as big as they felt. Maybe they were a virus examining the chemistry of its cell wall and believing it was studying the cosmos. Maybe the wall we kept hitting was not the limit of reality. It was the limit of the cell. And beyond the cell, the answers might exist. Simple, ordinary answers that require nothing more than a change of scale to understand. Answers that we would never reach, because we cannot leave the cell, because we do not know if we are in one.

Every question in this book might have an answer. Just not for us. Not at our scale. Not from inside whatever this is. The questions are real. The curiosity is real. The effort is real. And none of it may be capable of reaching beyond the walls of the thing we live in.

If that possibility is real, and nothing says it is not, then this entire book, every chapter, every argument, every theory examined and discarded, might be exactly what the title promised.

Pointless.

Not because asking is wrong. But because the answers, if they exist at all, might exist somewhere we cannot go, at a scale we cannot reach, in a reality we will never see. We might be the virus. The cell might be our universe. And the body might not know we are here.

Conclusion

This book asked questions. It answered none of them.

That was the point.

Each chapter began with a question that felt simple and ended in a place that felt unresolved. The origin of the universe. The nature of expansion. The possibility of something from nothing. The meaning of nothing itself. The source of matter. The need for answers. The illusion of understanding. The limits of science, logic, and language. The problem with asking why. The questions that cannot exist. The role of the observer. And the possibility that everything we think we know is the local knowledge of an organism too small to see the system it lives in.

None of these were answered because none of them can be answered. Not with current knowledge. Possibly not with any knowledge. Not because the questions are too hard. Because the questions operate at the boundary of what human cognition, human language, and human science can reach. They are not puzzles waiting for a smarter generation. They are walls built into the structure of inquiry itself.

Every attempt at an answer followed the same pattern. A theory was offered. The theory was coherent, well-reasoned, sometimes brilliant. And then the theory was shown to be an explanation, not an answer. A description of the problem in different language. A restatement of the mystery using more sophisticated tools. The mystery itself did not move. It sat exactly where it was, untouched, while the explanations piled up around it like scaffolding around a building that was never constructed.

Science offered mechanisms but not reasons. Religion offered narratives but not evidence. Philosophy offered frameworks but not conclusions. Logic offered structure but not completeness. Each discipline reached its limit and stopped. And at the edge of each discipline, the same silence.

The questions are real. The desire to answer them is real. The frustration of failing to answer them is real. But the answers, if they exist, do not appear to be accessible from where we stand. We are inside the system. We are made of the system. We are using tools built from the system to ask questions about the system. And the system does not appear to contain its own explanation.

Maybe it does. Maybe the answer to every question in this book is sitting in plain sight, obvious and simple, and we cannot see it because we are too small, too local, too embedded in the frame to recognize it. Maybe we are the virus in the cell, and the answer is the body, and the body is right there, all around us, and we cannot perceive it because our instruments point inward when they should point outward, and we do not know which direction outward is.

Or maybe there are no answers. Maybe the questions are artifacts of a brain that evolved to seek closure and cannot accept its absence. Maybe the universe does not owe us an explanation. Maybe reality has no obligation to make sense to an organism that has existed for an eyeblink of cosmic time on a small rock orbiting an ordinary star in an unremarkable corner of one galaxy among trillions.

Either way, the result is the same. The questions remain open. The mind remains unsatisfied. The book ends without resolution.

You started this book expecting answers. You are finishing it without any. And somewhere between the first page and this one, if the book did what it was meant to do, something shifted. Not toward an answer. Toward a recognition. That the limitation is not in the answer. It is in the question. That some questions do not lead anywhere. That the road the mind insists must continue might simply end. That the blank space at the edge of understanding is not a gap waiting to be filled. It might just be the edge.

This book was called The Pointless Book. And by now, you understand why. Not because the questions do not matter. They do. Not because curiosity is a flaw. It is the best thing about us. But because the questions, pursued honestly, lead to a place where the mind has nothing left to hold onto. No answer. No resolution. No closure. Just the questions themselves, sitting in the silence, unanswered and possibly unanswerable.

And yet.

You read the whole thing. You followed every chapter. You sat with every unanswered question. You did not stop. The mind that cannot tolerate the absence of answers kept going, page after page, through chapter after chapter of deliberate, sustained, unresolved tension. You did not get what you came for. And you kept reading anyway.

Which raises one final question. The only one this book will leave you with. The only one that matters.

Are questions without answers really pointless?

Elias Verdan

HEINRICH WILSON
H.W.
PUBLISHING
BOOKS WHICH REFUSE TO BE SILENT

www.ingramcontent.com/pod-product-compliance
Lightning Source LLC
LaVergne TN
LVHW022100200926
848481LV00003B/998